Representation in Scientific Practice

Representation in Scientific Practice

edited by Michael Lynch and Steve Woolgar

The MIT Press
Cambridge, Massachusetts
London, England

First MIT Press edition 1990

This book first appeared as a special issue of *Human Studies,* vol. 11, nos. 2–3 (April/July 1988). The essays by Françoise Bastide and Bruno Latour have been added for this edition.

Library of Congress Cataloging-in-Publication Data

Representation in scientific practice / edited by Michael Lynch and Steve Woolgar.—1st ed.
p. cm.
"First appeared as a special issue of Human Studies, vol. 11, nos. 2–3 (April/July 1988)."—T.p. verso.
Includes bibliographical references.
ISBN 0-262-62076-6
1. Scientific illustration. 2. Representation (Philosophy)
I. Lynch, Michael, 1948– . II. Woolgar, Steve.
Q222.R46 1990
502.2′2—dc20 90–5466
CIP

CONTENTS

Preface

All but two of the papers in this volume were published previously in a special issue of the journal, *Human Studies* (vol. 11), nos. 2–3, April/July, 1988. In the short time since the special issue was published, there has been a surge of interest among historians, philosophers, and sociologists in the diverse representational forms used in science. This is especially so for visual modes of representation, a topic that until recently was of interest to only a very few historians and sociologists of science. Because the papers in the special issue are more pertinent than ever to the emerging issues and debates on the topic, we decided that it would be worthwhile republishing the collection.

As stated in our introductory essay, the papers in this volume are diverse in several respects, and it is difficult to elaborate a "cover story" that holds for all of them. Although many of the papers focus on visual forms of representation—graphs, diagrams, photographs, and other instrumental displays—it would be inaccurate to say that this volume is solely about visual representation. Indeed, we would not want it to be about visual representation per se, since that would assign undue privilege to a particular representational form. If the studies in this volume agree on anything, it is that scientists compose and use particular representations in a contextually organized and contextually sensitive way. This, of course, is hardly unique to scientists' activities, as numerous ethnomethodological and constructivist studies of practical actions have demonstrated. A central tenet in ethnomethodology and in the sociology of scientific knowledge is that it is misleading to investigate language through an analysis of words and "meanings" isolated from the pragmatic situations in which they are used. This applies not only to words but just as forcefully to methodological rules and theoretical propositions. Although such formulations have their particular uses in scientific research, they are all too easily fetishized in academic treatments of scientific "ideas" and experimental "methodology." A kind of mysterious force and historical significance seems to accrue to rules and propositions when they are isolated from their pragmatic contexts. And, as debates in the philosophy of language continue to demonstrate, such fetishism gives rise to an intractable swarm of logical conundra.

Although we are pleased to see that philosophers, historians, and sociologists of science are now questioning the privilege traditionally assigned to the verbal statement or "proposition," and that they are devoting increased attention to visual forms of representation, we're less sanguine about some of the treatments being given to these phenomena. Specifically, we notice a tendency to transfer correspondence theories of representation from a propositional to a pictorial base. So, where the proposition was once held to be a word-picture of a logically closed system of facts (which could themselves, in principle, be decomposed into statements), a diagram or photograph is now treated as a "representation" of a theory and/or of "reality." Pictorial forms are then seen to have a particular potency for communicating "messages" about nature or indicating the content of "ideas." This, of course, is a very powerful view of representation, and one that continues to hold sway in analytic philosophy. The problem is that this view assumes a severely limited and *asocial* view of practical and communicative actions.

The studies in this volume endeavor in a variety of ways to show that the particular "representations" they discuss have little determinate meaning or logical force aside from the complex activities in which they are situated. Scientists compose and place representations within texts, data sets, files, and conversations, they juxtapose different forms of representation, and they use them in the course of a myriad of activities. Scientists do different "jobs" with them. Even for "highly realistic" diagrams and photographs, it does not do justice to the pragmatic and conventional features of these representational devices simply to describe the things they depict or the meanings they reflect. Such a naturalistic reading contravenes the basic policy of "anthropological strangeness" which informs our understanding of scientific representations. This policy requires us to put aside the view of representations as transparent images of objects and ideas, and to pay extraordinary attention to the distinctive surfaces upon which representations are inscribed and the translations they undergo when transferred from one activity to another. Rather than fetishizing representational forms, the papers in this volume elucidate the textual arrangements and discursive practices that produce and reproduce the mundane conditions for representational transparency.

To the papers from the *Human Studies* collection we have added an influential essay by Bruno Latour, "Drawing Things Together," and a detailed example of semiotic analysis by Françoise Bastide, "The Iconography of Scientific Texts: Principles of Analysis." Both papers had previously appeared, in French, in a special issue of *Culture Technique* (*numéro 14, June* 1985), entitled *Les 'Vues' de L'Esprit,* and co-edited by Bruno Latour and Jocelyn de Noblet. Latour's paper also appeared in English, under the title, "Visualization and Cognition: Thinking with Eyes and Hands," in *Knowledge and Society: Studies in the Sociology of Culture Past and Present,* Vol. 6, 1986:1–40). The paper in the present volume incorporates some minor changes from the earlier English version. We are happy to include it because of its comprehensive review of relevant work on visual representation and representational technologies. Latour not only covers a wide range of works in art history, literary studies, anthropology, and the history and sociology of science, he raises and elucidates an original set of themes. The paper's development of the theme of "inscription" (introduced in Latour and Woolgar's *Laboratory Life,* Sage, 1979), and its introduction of the concept of "immutable mobiles," provided a major impetus for research on visualization in science. Evidence of this influence can be found throughout the present volume. Latour initially developed the paper for a theme conference on Visualization et Connaissance, held at the Centre de Sociologie de l'Innovation de l'Ecole National Supérieure des Mines de Paris in December 1983. Latour's keynote paper was designed to articulate themes for discussion and debate in the sessions, and it succeeded marvelously at doing so. François Bastide, Karin Knorr-Cetina, John Law, and Michael Lynch were among the participants at the conference, and their papers for the present volume can be read as continuations of conversations that began in Paris in 1983.

Although Françoise Bastide's research on the semiotics of scientific illustrations has had considerable influence on social and historical studies of scientific texts, her work has only begun to be translated into English. Shortly after Bastide's untimely death in 1988, Greg Myers completed a translation of her paper on the iconography of scientific texts. Because we thought that it would add immeasurably to the collection of papers in this volume, we arranged with Myers and Bastide's estate to include it. Both Bastide's and Latour's papers exemplify a distinctive approach to representation that creatively synthesizes semiotic, post-structuralist, and social-constructivist initiatives. The cross-references between them and the other papers in the volume contribute to an internal dialogue that helps to specify common themes and critical differences.

We are very grateful to Kluwer Academic Publishers, the publisher of *Human Studies,* and especially to George Psathas, editor of the journal, for permitting us to republish the papers from the special issue. We also would like to thank JAI Press, Inc., for permission to reprint Bruno Latour's paper, and Françoise Bastide's estate for permitting our publication of the English translation of her paper.

Introduction: Sociological orientations to representational practice in science

MICHAEL LYNCH
Department of Sociology, Boston University, Boston, MA 02215, USA
STEVE WOOLGAR
Department of Human Sciences, Brunel University, Uxbridge, Middlesex UB8 3PH, UK

1. Introduction

Contributors to this Special Issue were asked to investigate the topic of "representational practice in science." We did not require that the studies adopt a particular theory of representation, or follow a specific methodological approach. Nor did we specify a preference for the fields of scientific practice to be investigated. Contributors were asked to consider any of the varieties of representational devices used in science: graphs, diagrams, equations, models, photographs, instrumental inscriptions, written reports, computer programs, laboratory conversations, and hybrid forms of these. Studies on the organization, interconnectedness, and use of such devices were to be related to general issues in the sociology and philosophy of science.

Given this open-ended agenda, it is not surprising that this Special Issue contains a diverse collection of papers. With the exception of Paul Tibbetts' essay – a critical discussion on the sociology of scientific representation which helps to specify the context of inquiry for the other studies – the papers examine particular texts and activities. They investigate a number of different kinds of representation: visual and graphic documents of laboratory data, textual figures, biographical accounts of scientific "problems," instructions for novices, and "artificially intelligent" programs. While there are some "family resemblances"[1] between the themes discussed in the papers, their very diversity serves to demonstrate what we take to be a key "finding" about representation in science: the heterogeneity of representational order. This includes more than the evident diversity in kinds of representational device, as it covers the theoretic principles and functions of representation as well: resemblance, symbolic reference, similitude, abstraction, exemplification, expression.

An emphasis on the heterogeneity and discontinuity of representational formats, and on the local and contextual basis of their production, has become a familiar refrain in 'postmodernist' critiques of traditional views on science and language (Derrida, 1970; Foucault, 1970; Lyotard, 1984). The word-object relationship, once the paradigm of representation, is displaced by lateral, syntagmatic, and reflexive relations between communicational 'elements' in seemingly anarchistic fields. Universal or essential meanings have been divested of their genealogical authority over the 'superficial' organizations and localized uses of textual order. But, it is evident that the papers in this issue focus on specific cases of representational activity in science, and nothing so global as the purported 'condition of science in modern society'.

The papers in this collection do not share a single analytic perspective (such as semiotics, ethnomethodology, speech-act theory, constructivism, deconstruction, hermeneutics, etc.), but they do share a current situation of inquiry in social studies of science. Because readers of this journal may not be familiar with research in this field, we will give a brief "historicized" account of the development of a sociological interest in scientists' representational practices.

2. A sociological interest in the 'contents' of scientific knowledge

Until the 1970s sociologists had little incentive to investigate such matters as how scientists construct models, enact experimental runs, design and interpret data displays such as bubble-chamber tracks or autoradiographs, publically report upon methods and findings, and assign credit for discoveries. These were assumed by most sociologists to be the province of epistemology, or, on a more specific level, of the specialized sciences themselves.

Sociologists concerned themselves with "institutional" aspects of science abstracted from specific research contexts: ethical norms, systems of rewards, status differentials, community configurations, role requirements, and stereotypic personalities. "Social" explanations were given to experimental conduct and findings only under conditions where they were found to be erroneous or fraudulent.

This situation changed when historians and philosophers of science began to use more "sociological" forms of explanation when accounting for theory acceptance and theory change, the emergence of specialties, and the persistence and resolution of controversies. Writers such as Fleck, Kuhn, Polanyi, Hanson, Holton, Lakatos, and Feyerabend emphasized the importance of "agreement" in communities of scientists on matters of fact and procedure, and of efforts to enlist such agreement through persuasive appeals. Sociologists eventually extended and elaborated upon such leads, and developed several loosely linked "programs" of study.

The most well known of these is the "strong programme" in the sociology of scientific knowledge (Bloor, 1976; Barnes, 1974, 1977; Barnes and Shapin, 1979). Taking initiatives from Wittgenstein's later writings on philosophy of language, Mary Douglas' cognitive anthropology, and Kuhn's (1962) and Hesse's (1974) histories and philosophies of science, Bloor and Barnes developed several research policies which they and their colleagues exemplified with socio-historical case studies.

A basic tenet of the "strong programme" is to devise sociological explanations of all scientific "beliefs," regardless of the truth or falsity eventually assigned to such beliefs. So, for instance, the development of electromagnetic theory or of the model for the biochemical structure of DNA should not be explained simply as "correct" beliefs reflecting the order of nature and the disinterested performance of exact procedures, while the Piltdown hoax or the short-lived N-ray "affair" are given 'social' forms of explanation (i.e., by citing intense competition between rival researchers, nationalist commitments, incorrigible assumptions, and closed epistemic communities). For Barnes and Bloor, and those influenced by their programmatic writings, scientific history is driven by tangible social interests and linkages, and it is inappropriate for sociologists to make judgments about the correctness of a belief before determining whether it warrants sociological explanation.

2.1 *What are 'the contents' of scientific knowledge?*

The "strong programme" treated the 'contents' of scientific knowledge as reified products of: historical and contemporary controversies; partisan genealogies of scientific facts; and negotiations over credit for discoveries. The 'technical contents' of discovery, experimentation, replication, argumentation, and representation now made up a roster of 'sociological' topics, to be studied as *situated* processes of knowledge-production and not exclusively as methodological and epistemological concerns. But, while many of us agreed that the 'technical contents' of scientific investigations were fair game for sociologists, what exactly such 'contents' might be and how they should best be studied was far from established. Socio-historical arguments on how scientists' theoretical commitments reflected social "interests" were suggestive but begged many questions on the nature of such "interests" and how they work themselves into research settings (Woolgar, 1981), and, as Paul Tibbetts argues in his paper, Bloor's and Barnes' constructivist arguments are reflexively undermined by incongruent causal and realist assumptions (see also Laudan, 1981; and Woolgar, 1983).

As a way of getting closer to the 'technical contents' they sought to comprehend, some sociologists of science began to study contemporaneous cases of scientific practice. Many of them adopted themes and methods previously worked out in ethnomethodological and other interactionist approaches to practical action in settings other than science.[2] In some instances, sociologists and anthropologists proceeded ethnographically by spending extended periods of time in one or a few laboratories, conducting extensive interviews, collecting and analyzing laboratory documents, and tape recording informal "shop talk" (Latour and Woolgar, 1986; Knorr-Cetina, 1981; Lynch, 1985a; Collins, 1985; Law and Williams, 1982; Law, 1986). Other studies were more exclusive in their "textual" focus, examining such specific documents as discovery reports (Woolgar, 1976; 1980), published articles (Gusfield, 1976; Bazerman, 1981; Morrison, 1988), and Nobel Prize acceptance speeches (Mulkay, 1984). Both the ethnographic and the textual studies focused on the 'rhetoric' in scientific writing, the collaborative performance of experiments, and informal agreements and disagreements over the sense and import of laboratory data.

According to many of these studies, the "contents" of scientific research are documents or, in Latour and Woolgar's (1986) terms, "inscriptions." Manifestly, what scientists laboriously piece together, pick up in their hands, measure, show to one another, argue about, and circulate to others in their communities are not "natural objects" independent of cultural processes and literary forms. They are extracts, "tissue cultures," and residues impressed within graphic matrices; ordered, shaped, and filtered samples; carefully aligned photographic traces and chart recordings; and verbal accounts.[3] These are the proximal "things" taken into the laboratory and circulated in print, and they are a rich repository of "social" actions. In order to see how such documentary resources and products can be relevant to sociological considerations, it is necessary to disclose how they are more than simply representations of natural order.

3. Serial orders of representations

There is no necessity to take an "anti-realist" (cf. Hacking, 1983: 21ff.) position in order to grasp that graphic displays and other representations are not simply pictures of natural objects. Whether or not one believes in the reality of the entities and theoretical relationships made visibly present in, e.g., electron micrographs or autoradiographs of systematically prepared tissue, it is possible to see that other, equiprimordial, representational orders are created and sustained through scientists' use of such documents. Representations can represent *other* representations in complex socio-technical networks: the sense conveyed by a picture may derive as much from a spatio-temporal order of other representations as from its resemblance or symbolization of some external object. Relationships *between* representational objects and expressions are of particular interest for any effort to reveal the "social" organization of technical work in science.

Several of the papers in this Special Issue elucidate "lateral" or "serial" relations between representational products of scientific work: Amann and Knorr-Cetina describe how visual evidences are "fixated" in a succession and collation of referential formats; Woolgar discusses an unfolding "trace" on a chart recording and the variable sense that is made of it in a laboratory project; Lynch speaks of "rendering practices" through which specimen materials are successively transformed into mathematized icons; and Myers shows how different illustrative formats are used progressively to link a sense of "natural" reality to "abstract" theoretical relationships.

3.1 Directionality

Serial organizations of representations are "progressive." The transformations of, e.g., living brain tissue into graphic exhibits of anatomical constituents or biochemical fractions create sensual properties. Boundaries, discriminations, spatial orientations and temporal series are exposed and framed on graphic surfaces. The phenomena under study become more visible, stable, and measurable – they more evidently "fit" (even if they do not unambiguously support) what is claimed about their structure. The question then is, "Is such 'progress' irreversible?" Is an ordered series of representations an unbroken chain linking natural material to underlying theoretical structure? Is a series of representations to be viewed as a series of pictures of a same thing, where differences are governed by the plenum of possible phases, emissions, or perspectives under which some thing can be witnessed? We cannot answer these questions in a general way, although we can note that regardless of what one assumes about the reality of the represented objects, there are many circumstances under which "chains" of representations may be reversed or broken into discontinuous fragments. Close examination of the relations between one representation and another reveals "transferences" of graphic and other materials across a series of disjointed surfaces: a crafting of resemblances.

3.2 Decalcomanie[4]

> Resemblance makes a unique assertion, always the same: This thing, that thing, yet another thing is something else. Similitude multiplies different affirmations, which dance together, tilting and tumbling over one another. (Foucault, 1982:46)

For Foucault (1970; 1982), resemblance presumes a primary referent, an "original" that any "copy" renders in a partial and imperfect way. With similitude there is no ultimate reference point. Representational relations are symmetric and can be reversed and extended without limit. Each item in a network of similitudes is a "graphism that resembles only itself" (1982:48). Similarity relations are exposed in the way diverse texts are brought together. Juxtaposition, and not the inherent characteristics of an "original," provides the material basis for the disclosure of similitudes. An art which makes similitude explicit is not simply different from representation by resemblance. It exposes a common ground that is hidden in representational realism, and thereby claims primacy for similitude over resemblance. Resemblance is a product of "illusionist" technique (Gombrich, 1960), while similitude is an inescapable resource in any textual representation. Hence, to analyze representation is to expose the conjurer's tricks through which chains and networks of similitude are laboriously built-up and then "forgotten" in the presumptive adequacy of their reference to an "original."

How does this apply to scientific representation? For science, as typically understood, both resemblance and similitude are inadequate. "Mere" metaphor, similarity, or surface resemblance are to be discounted in favor of "deep," "genetic," or "mathematical" reconstructions of a phenomenon's organization. It is not enough to represent the object; it must be *penetrated* by theory and opened-up to an active manipulation of its principles of organization. But, scientists do describe and depict. They utilize the full range of literary devices and artistic conventions available to them (Edgerton, 1976; Alpers, 1983).

Several of the studies in this Special Issue show that the differences between drawings, instrumental inscriptions, autoradiographs, etc. are not incidental. Each type of representation provides distinctive formats, referential uses, and iterable and analyzable traces. If it is a scientist's task to plumb the "depths" of a phenomenon, available representations provide the material with which such "plumbing" is visibly constructed.

A recommendation we can give for reading some of the papers in this collection is to consider serial, "directional," relations between representations, and differences in the abstracted or naturalistic form of representations, to be relations between technical products in a work process. The "direction" is not a movement away from or toward an originary reality, but a movement of an assembly line.[5]

4. Textual bricolage

Lévi-Strauss (1966) opposes the bricoleur to the engineer, while Garfinkel identifies bricolage as an indispensible set of practices for making experiments work (Garfinkel et al., 1981; also see Amann and Knorr-Cetina's essay in this issue). The bricoleur improvises with a mixed bag of tools to adapt to the task at hand, while Lévi-Strauss's engineer devises a plan, assembles specialized tools and materials, and then articulates the specifications in the plan. Close examination of just how scientists articulate experimental designs and use specialized equipment turns the opposition between bricoleur and engineer into a genealogy: bricolage is at the heart of the work of sustaining a plan and remedying its provisions in light of unanticipated contingencies.

Much has been said about contingencies in laboratory work and their critical implications for any theory of scientific practice (see Collins, 1985; and Garfinkel et al., forthcoming) for detailed accounts on this issue). One such implication has been to shift attention from experimental designs and formulae to the embodied performance of experimentation and the situated use of formulae. The praxis of instrumental calibration and use, and the social interactions through which equipment is assessed and results are negotiated, prove to be rich topics for sociological investigation that were obscured when attention was focused strictly on principles of experimental design and confirmation.

Several of the papers in this Special Issue contribute to our understanding of scientists' bricolage. Each paper does so by locating settings where discrepancies between representations of practice and the practices using (and composing) such representations are especially pronounced. The import of such discrepancy is not simply to question the authority of standard representations of practice, but to reveal previously ignored constituents of scientific work.

Amerine and Bilmes provide us with observations of grade-school children attempting to perform classroom demonstrations designed to instruct them about scientific principles. The kids prove to be apt improvisors, discovering and exploiting contingent ways to perform the exercises. Although, in one sense, what the kids do with the experimental apparatus creates tangential issues which undermine the "lesson" to be demonstrated, in another sense they discover ways of contending with incomplete instructions and unforeseen events. Law and Lynch utilize the simple (pre-scientific?) example of amateur bird watching to discuss representations of bird species in field manuals, and problems of "reading" such manuals in specific field situations. Like the kids in Amerine and Bilmes' paper, birdwatchers experience troubles and devise remedies which are, perhaps, emblematic of more sophisticated scientific observations. Suchman, on the other hand, examines the epitome of professionally planned action – Artificial Intelligence research – and describes how the rational actions attributed to "intelligent artifacts" are sustained through tacit modes of accommodation in human-machine interaction.

Yearley's paper takes a somewhat different line of attack, but also treats the discrepancy between generalized formulations of scientific procedure and localized performances of such procedures. Yearley collects and analyzes accounts by scientists on the "problems" they have been studying, and notes some of the variable ways they formulate "biographies" of their projects. During interviews, scientists rapidly shifted their versions of the objectives for problems and projects, sometimes mentioning opportunistic modifications of research objectives as distinct funding and other career opportunities emerged, while at other times presenting the research as pursuit of technical problems which seemingly were present from the outset. Yearley suggests that the latter sort of account is more likely to be given at conference presentations and similar settings.

Textual bricolage – the improvisory work of assembling and reading accounts of research practices and results – is revealed by closely studying how textual and other representations are used in the course of specific inquiries. It cannot be disclosed by inspecting visual or other representations by themselves, and yet it can be viewed as a way of "reading" such textual materials. A line traced by an instrument on a chart recording, can be read in a variety of ways: its features can be treated as evidence of any number of worldly events, or of malfunctioning in the complex of instruments. How the display is read depends upon scientists' efforts to insert the document into the complex socio-technical relevancies of day-to-day investigation: who assembled the equipment, how it worked the last time it was used, what sorts of things have gone wrong or could go wrong with the apparatus, what sorts of proximal and distal events can the recording instrumentation "pick up," etc. (and "etc." is a key feature of this configuration). In Woolgar's and in Amann and Knorr-Cetina's papers, tape recordings of researchers' shop talk are used as a basis for analyzing how scientists work-through such contingent possibilities. While a trace on a chart recording may be "iterable," its local "reading" often turns out not to be stable or predictable from the document alone.[6] Stability (or in Amann and Knorr-Cetina's terms, "fixation") of documentary evidence occurs as a consequence of a temporal succession of practices in a lively social-interactional setting.

5. From representation to representations

If, for the early Wittgenstein, the traditional aim of philosophy is to clarify propositions and prevent us from being led astray by the misleading appearances of everyday language, sociology, as we understand it, has quite a different task. The logical analysis of representational propositions, or of the concept of representation itself, provides philosophers with a generalized sense of what it means "to represent." Sociologists, on the other hand, have no need to *clarify* everyday language, insofar as "everyday" or "ordinary" actions are the primary subject of empirical study. Rather than being a source of inadequate or misleading appearances, everyday actions for sociologists constitute orders to be investigated in their own right.

It may, of course, be presumptuous to propose an empirical study of anything, and especially to propose an empirical study of scientific activities. But, when we say that the studies in this special issue treat representation empirically, we refer to their particularistic case-study approach, and not to any inherent basis for being more "grounded" or less speculative than philosophers' reflections on the subject. As earlier noted, the studies focus on specific contexts in which representations are produced and displayed, and not on the generalized concept of representation. The advantage of such an approach is that organized uses of representations which were not previously imagined can be "stumbled upon" in a specific case study. We do not claim that such an approach guarantees that the descriptions offered in a particular study completely or correctly reproduce 'what actually occurred,' or that such descriptions are unprejudiced. Instead of asking, "what do we mean, in various contexts, by 'representation'?"; the studies begin by asking, "What do the participants, *in this case*, treat as representation?"

5.1 Reflexivity

Representation is an explicit issue in science (as in selected other activities like art) – it is addressed in and as the activity. Scientific representation is a topic of an immense literature; and proposals about how it is done penetrate the self-understanding of avowedly scientific disciplines like sociology. As sociologists, we are already stuck with the legacy of previous wisdom about scientific representation, and we cannot escape from this even when we treat what other scientists do as a topic. There are at least two implications of this:

1. The organization, sense, value, and adequacy of any representation is "reflexive" to the settings in which it is constituted and used. This ethnomethodological version of reflexivity asserts that, for instance: the sense and analytic structure of a

textual figure depends upon the figure's use in the course of the text's reading (Morrison, 1987); the sense and import of a "pulse" on an oscillograph screen is reflexive to the local, interactionally organized, work of "extracting" the pulse from the practical contingencies of a series of observational runs (Garfinkel et al., 1981); the organization and rigor of a mathematical proof is reflexive to mathematicians' practices of demonstrating to competent colleagues how the proof is "followable" (Livingston, 1986).

"Reflexivity" in this usage means, not self-referential nor reflective awareness of representational practice, but the inseparability of a "theory" of representation from the heterogeneous social contexts in which representations are composed and used. To study representations 'sociologically,' thus means to come to terms with the socially distributed competencies which establish the theoretic sense and import of any representational device. Such indigenous "theories" of representation include more than, for instance, the explicit instructions, justifications, etc. formulated in texts on scientific methodology. As detailed in the studies in this Special Issue, they are expressed in the way representations are put together, discussed in casual conversations, selected and placed in textual arguments, and 'read' in light of the vicissitudes of a practical situation.

2. It also can be noted that the studies in this collection *use* as well as analyze representations. Representative documents and excerpted passages attributed to "scientists" are selected and reproduced as objects or evidence for "sociological" claims. Not much is said about this in the papers, as they mainly deal with how the features of such representations are reflexive to *other* textual uses and contexts of inquiry. What can we make of this?

One thing we will *not* claim is that by studying scientists' representational practices we learn to use representations in more rigorous, convincing, defensible ways. Claims about the "self-exemplying" (Merton, 1978) use of findings in the sociology of science are overblown, since they presume isomorphism between sociology and various other sciences studied, and worse, their attempts to emulate 'science' suppress any more critical understanding of the phenomena under study.

It should be expected that, should one be inclined to investigate the matter, that representations used by sociologists of science are selected, arranged in series, identified with captions, etc. in order to make the case for various claims about 'objects' such as: scientists' practices, a textual argument, kids' efforts to follow instructions, etc. Further, a hypothetical study of the papers in this Special Issue might reveal that many of the analytic claims they make about what they study depend upon unmentioned representational "devices." Such "devices" act as a foundation for argument no less significant than any logical relation between the authors' claims and what scientists "really do."

The significance we attach to this observation depends crucially on the extent to which we intend our study of representational practice in science to be critical. To claim that our investigations reveal deficiencies in representational practice in science would be to assume a correspondence between argument and object as our ideal representational aim. In that event our own sociological argument would be open to the same charge of deficiency. At the same time, the very idea of deficiency implies the availability of 'objects' which are somehow free of representation. On the contrary, our position is that representations and objects are inextricably interconnected; that objects can only be 'known' through representation. Criticism necessarily involves competition between representations, not between representation and an 'actual object.' On the other hand, our pretensions to critical neutrality are also problematic. The normal conventions of written academic discourse seem to require the frequent use of a distinction between representation and object, and this permits our comments about scientific practice to be read as being critical, if only by innuendo (cf. Woolgar and Pawluch, 1985).

It is not yet clear to what extent it is possible to escape the constraints of conventional academic discourse (for attempts to examine the consequences of reflexivity through the exploration of 'new literary forms,' see Mulkay, 1985; Woolgar, 1988). The various case studies in the Special Issue do not attempt to transcend conventional and polemical usage of visual and other representational devices. However, they do attempt to examine such conventional usage in restricted domains by examining particular contexts in which representations are used as 'scientific' resources. Such close attention to specific instances, whether of scientists' representational practice or of sociologists' representations of scientists' representations, can provide a first step for re-examining our received textual wisdom.

Notes

1. We use Wittgenstein's (1953) term "family resemblances" to speak of lateral connections between one study and another and not a "deep" connection of all of the studies to an abstracted "core idea."
2. Ethnomethodology is a discipline which studies how social order is produced through local interaction. Ethnomethodologists examine ordinary conversations and other institutionalized activities to explicate the constitutive "work" through which routine actions, recognizable social identities, and collective events are accomplished (see Garfinkel, 1967; Cicourel, 1964; Heritage, 1984). Themes from Garfinkel's and Cicourel's critiques of standard sociological methods, and from ethnomethodological studies on topics such as the institutionalized management of social deviance influenced many of the studies discussed above. In an independent development, Garfinkel and some of his students turned their attention in the late 1970s to natural scientists' and mathematicians' practices and developed a distinctive program of study (see Garfinkel et al., 1981; Lynch et al., 1983; Garfinkel et al., forthcoming; and Livingston, 1986).
3. See Latour and De Noblet (1985), Latour (1986), Lynch (1985b), Gilbert and Mulkay (1984), Pinch (1985), Shapin (1984), Rudwick (1976), Gooding (1986), for studies on the history and contemporaneous uses of visual and verbal formats for witnessing and presenting evidentiary accounts of scientific phenomena.
4. Foucault (1982) uses the title of Magritte's (1966) painting to identify a general property of representation. In Magritte's *Decalcomanie* a man in a bowler hat occupies the left half of the picture and stands silhouetted against a blue sky, his back to the viewer. A red curtain blocks the view of the sky to the man's right. Another silhouette with exactly the same outline as the first occupies the right half of the picture, and is superimposed against the red curtain. This silhouette displays the "scene" of sea and open sky presumably blocked to view by the man's body and his bowler hat.

   For Foucault, this picture displays a "displacement and exchange of similar elements, but by no means mimetic reproduction" (page 46). For Foucault (1982:62, note 3), *De Calcomanie* identifies a multifaceted phenomenon:

> *Decalcomanie* means transference, transferency, or decal; it is also a painterly technique (often mentioned by Breton) in which pigment is transferred from one side of a painted surface to another by folding over the canvas. Finally, *decalcomanie* refers to a species of madness bound up with the idea of shifting identities.

5. Yoxen (1987) notes that the progression toward more schematic, geometric renderings described in Lynch's (1985b) analysis of electron microscopy is reversed in the case of ultrasound imaging technology. There, the development is toward progressively more detailed depiction of form. The contrast between the two cases indicates that it would be inappropriate to assume that scientific technique transforms "natural" into "artificial" form; "natural" form can itself be a technical product.
6. This is not to suggest that we should ignore Derrida's recommendations and subsume the properties of writing to speech. As Latour (1986) suggests, printed texts and evidentiary documents are "immutable mobiles" – fixed renderings that can be widely circulated through dispersed research sites, thereby providing a stable literary basis for a disciplinary network. But, as stated above and elaborated in several of the papers in this issue, while printed (or otherwise inscribed) figures can be temporally and spatially removed from their original sites, when we inspect the work at such sites we find that documents can be crumpled up and thrown away, lines and words revised, textures enhanced or toned down. "Immutability" is not guaranteed by the properties of printed media, nor is "mobility" achieved except for locally composed and selected documents.

References

Alpers, S. (1983). *The art of describing*. Chicago: University of Chicago Press.

Barnes, B. (1974). *Scientific knowledge and sociological theory*. London: Routledge and Kegan Paul.

Barnes, B. (1977). *Interests and the growth of knowledge*. London: Routledge and Kegan Paul.

Barnes, B., and Shapin, S., Eds. (1979). *Natural order: Historical studies of scientific culture*. London: Sage.

Bazerman, C. (1981). What written knowledge does: Three examples of academic discourse. *Philosophy of the Social Sciences* 11(3):361–387.

Bloor, D. (1976). *Knowledge and social imagery*. London: Routledge and Kegan Paul.

Cicourel, A. (1964). *Method and measurement in sociology.* New York: The Free Press.

Collins, H. (1985). *Changing order: Replication and induction in scientific practice.* London and Beverly Hills: Sage.

Derrida, J. (1970). Structure, sign, and play in the discourse of the human sciences. In R. Macksey and E. Donta (Eds.), *The structuralist controversy: The language of criticism and the sciences of man.* Baltimore: Johns Hopkins University Press.

Edgerton, S. (1976). *The renaissance discovery of linear perspective.* New York: Harper and Row.

Foucault, M. (1970). *The order of things.* Trans. A. Sheridan. New York: Pantheon.

Foucault, M. (1982). *This is not a pipe.* Trans. J. Harkness. Berkeley: University of California Press.

Garfinkel, H. (1967). *Studies in ethnomethodology.* Englewood Cliffs: Prentice Hall.

Garfinkel, H., Livingston, E., Lynch, M., and Robillard, B. (Forthcoming). Respecifying the natural sciences as discovering sciences of practical action: (I and II) Doing so ethnographically by administering a schedule of contingencies in discussions with laboratory scientists and by hanging around their laboratories. In D. Zimmerman and D. Boden (Eds.), *Talk and Social Structure.*

Garfinkel, H., Lynch, M., and Livingston, E. (1981). The work of a discovering science construed with materials from the optically discovered pulsar. *Philosophy of the Social Sciences* 11(2):131–158.

Gombrich, E.H. (1960). *Art and illusion: A study in the psychology of pictorial representation.* Princeton: Princeton University Press.

Gooding, D. (1986). How do scientists reach agreement about novel observations? *Studies in History and Philosophy of Science* 17.

Gusfield, J. (1976). The literary rhetoric of science: Comedy and pathos in drinking-driver research. *American Sociological Review* 41:16–34.

Hacking, I. (1983). *Representing and intervening: Introductory topics in the philosophy of natural science.* Cambridge: Cambridge University Press.

Heritage, J. (1984). *Garfinkel and ethnomethodology*. Oxford: Polity Press.

Hesse, M. (1974). *The structure of scientific inference.* Berkeley: University of California Press.

Knorr-Cetina, K. (1981). *The Manufacture of knowledge: An essay on the constructivist and contextual nature of science.* Oxford: Pergamon Press.

Kuhn, T. (1962). *The structure of scientific revolutions.* Chicago: University of Chicago Press.

Latour, B. (1986). Visualisation and cognition: Thinking with eyes and hands. *Knowledge and Society: Studies in the Sociology of Culture Past and Present* 6:1–40.

Latour, B., and De Noblet, J., Eds. (1985). *Les "vues" de l'esprit.* Special Issue of *Culture Technique* 14 (June).

Latour, B., and Woolgar, S. (1986). Laboratory life: The construction

of scientific facts. 2nd edition. Princeton: Princeton University Press.
Laudan, L. (1981). The pseudo-science of science? *Philosophy of the Social Sciences* 11(2):173–198.
Law, J. (1986). On power and its tactics: A view from the sociology of science. *The Sociological Review* 34(1).
Law, J., and Williams, J. (1982). Putting facts together: A study of scientific persuasion. *Social Studies of Science* 12(4):535–558.
Lévi-Strauss, C. (1966). *The savage mind.* Chicago: University of Chicago Press.
Livingston, E. (1986). *The ethnomethodological foundations of mathematics.* London: Routledge and Kegan Paul.
Lynch, M. (1985a). *Art and artifact in laboratory science: A study of shop work and shop talk in a research laboratory.* London: Routledge and Kegan Paul.
Lynch, M. (1985b). Discipline and the material form of images: An analysis of scientific visibility. *Social Studies of Science* 15:37–66.
Lynch, M., Livingston, E., and Garfinkel, H. (1983). Temporal order in laboratory work. In K. Knorr-Cetina and M. Mulkay (Eds.), *Science observed: Perspectives on the social study of science.* London and Beverly Hills: Sage.
Lyotard, J. (1984). *The postmodern condition: A report on knowledge.* Trans. G. Bennington and B. Massumi. Minneapolis: University of Minnesota Press.
Merton, R. (1978). The sociology of science: An episodic memoir. In R. Merton and J. Gaston (Eds.), *The sociology of science in Europe.* Carbondale: Southern Illinois University Press.
Morrison, K. (1988). Some researchable recurrences in science and situated science inquiry. In D. Helm et al. (Eds.), *The interactional order.* New York: Irvington.
Mulkay, M. (1984). The ultimate compliment: A sociological analysis of ceremonial discourse. *Sociology* 18:531–549.
Mulkay, M. (1985). *The word and the world: Explorations in the form of sociological analysis.* London: Allen and Unwin.
O'Neill, J. (1981). The literary production of natural and social science inquiry. *Canadian Journal of Sociology* 6:105–120.
Pinch, T. (1985). Towards an analysis of scientific observation: The externality and evidential significance of observation reports in physics. *Social Studies of Science* 15:1–36.
Rudwick, M. (1976). The emergence of a visual language for geological science 1760–1840. *History of Science* 14:148–195.
Shapin, S. (1984). Pump and circumstance: Robert Boyle's literary technology. *Social Studies of Science* 14:481–521.
Wittgenstein, L. (1953). *Philosophical investigations.* Trans. G.E.M. Anscombe. New York: Macmillan.
Woolgar, S. (1976). Writing an intellectual history of scientific development: The use of discovery accounts. *Social Studies of Science* 6:395–422.

Woolgar, S. (1980). Discovery: Logic and sequence in a scientific text. In K. Knorr-Cetina, R. Krohn and R. Whitley (Eds.), *The social process of scientific investigation, sociology of the sciences yearbook*, Vol. 4. Dordrecht: Reidel.

Woolgar, S. (1981). Interests and explanation in the social study of science. *Social Studies of Science* 11:365–394.

Woolgar, S. (1983). Irony in the social study of science. In K. Knorr-Cetina and M. Mulkay (Eds.), *Science observed: Perspectives on the social study of science*. London and Beverly Hills: Sage.

Woolgar, S., Ed. (1988). *Knowledge and reflexivity: New Frontiers in the sociology of knowledge*. London: Sage.

Woolgar, S., and Pawluch, D. (1985). Ontological gerrymandering: The anatomy of social problems explanations. *Social Problems* 32:214–227.

Yoxen, E. (1987). Seeing with sound: A study of the development of medical images. In W. Bijker, T. Hughes and T. Pinch (Eds.), *The social construction of technological systems: New directions in the sociology and history of technology*. Cambridge, MA: MIT Press.

Drawing things together

BRUNO LATOUR
Centre de Sociologie, Ecole des Mines, Paris

1. Putting visualization and cognition into focus

It would be nice to be able to define what is specific to our modern scientific culture. It would be still nicer to find the most economical explanation (which might not be the most economic one) of its origins and special characteristics. To arrive at a parsimonious explanation it is best not to appeal to universal traits of nature. Hypotheses about changes in the mind or human consciousness, in the structure of the brain, in social relations, in "mentalités," or in the economic infrastructure which are posited to explain the emergence of science or its present achievements are simply too grandiose, not to say hagiographic, in most cases and plainly racist in more than a few others. Occam's razor should cut these explanations short. No "new man" suddenly emerged sometime in the sixteenth century, and there are no mutants with larger brains working inside modern laboratories who can think differently from the rest of us. The idea that a more rational mind or a more constraining scientific method emerged from darkness and chaos is too complicated a hypothesis.

It seems to me that the first step toward a convincing explanation is to adopt this a priori position. It clears the field of study of any single distinction between prescientific and scientific cultures, minds, methods, or societies. As Jack Goody points out, the "grand dichotomy" with its self-righteous certainty should be replaced by

An earlier version of this article was published under the title "Visualization and Cognition: Thinking with Eyes and Hands," in *Knowledge and Society: Studies in the Sociology of Culture Past and Present,* vol. 6 (1986), pp. 1–40. We thank JAI Press for permission to reprint the article here.

many *uncertain* and *unexpected* divides (Goody, 1977). This negative first move frees us from positive answers that strain credulity.[1] All such dichotomous distinctions can be convincing only as long as they are enforced by a strong asymmetrical bias that treats the two sides of the divide or border very differently. As soon as this prejudice loses hold, cognitive abilities jump in all directions: sorcerers become Popperian falsificationists; scientists become naive believers; engineers become standard "bricoleurs"; as to the tinkerers, they may seem quite rational (Knorr, 1981; Augé, 1975). These quick reversals prove that the divide between prescientific and scientific culture is merely a border—like that between Tijuana and San Diego. It is enforced arbitrarily by police and bureaucrats, but it does not represent any natural boundary. Useful for teaching, polemics, commencement addresses, these "great divides" do not provide any explanation, but on the contrary are the things to be explained (Latour, 1983).

There are, however, good reasons why these dichotomies, though constantly disproved, are tenaciously maintained, or why the gap between the two terms, instead of narrowing, may even widen. The relativistic position reached by taking the first step I propose, and giving up grand dichotomies, looks ludicrous because of the enormous consequences of science. One cannot equate the "intellectual" described by Goody (1977, ch. 2) and Galileo in his study; the folk knowledge of medicinal herbs and the National Institute of Health; the careful procedure of corpse interrogation in the Ivory Coast and the careful planning of DNA probes in a California laboratory; the storytelling of origin myths somewhere in the South African bush and the Big Bang theory; the hesitant calculations of a four-year-old in Piaget's laboratory and the calculation of a winner of the Field Medal; the abacus and the new super-computer Cray II. The differences in the *effects* of science and technology are so enormous that it seems absurd not to look for enormous causes. Thus, even if scholars are dissatisfied with these extravagant causes, even if they admit they are arbitrarily defined, falsified by daily experience and often contradictory, they prefer to maintain them in order to avoid the absurd consequences of relativism. Particle physics must be radically different in some way from folk botany; we do not know how, but as a stop-gap solution the idea of rationality is better than nothing (Hollis and Lukes, 1982).

We have to steer a course that can lead us out of a simple relativism and, by positing a few, simple, empirically verifiable causes, can account for the enormous differences in effects that everyone knows are real. We need to keep the scale of the effects but seek more mundane explanations than that of a great divide in human consciousness.

But here we run into another preliminary problem. How mundane is mundane? When people back away from mental causes, it usually means they find their delight in material ones. Gigantic changes in the capitalist mode of production, by means of many "reflections," "distortions," and "mediations," influence the ways of proving, arguing and believing. "Materialist" explanations often refer to deeply entrenched phenomena, of which science is a superstructure (Sohn-Rethel, 1978). The net result of this strategy is that nothing is empirically verifiable since there is a yawning gap between general economic trends and the fine details of cognitive innovations. Worst of all, in order to explain science we have to kneel before one specific science, that of economics. So, ironically, many "materialist" accounts of the emergence of science are in no way material since they ignore the precise practice and craftsmanship of knowing and hide from scrutiny the omniscient economic historian.

It seems to me that the only way to escape the simplistic relativist position is to avoid both "materialist" and "mentalist" explanations at all costs and to look instead for more parsimonious accounts, which are empirical through and through, and yet able to explain the vast effects of science and technology.

It seems to me that the most powerful explanations, that is, those that generate the most out of the least, are the ones that take writing and imaging craftsmanship into account. They are both material and mundane, since they are so practical, so modest, so pervasive, so close to the hands and the eyes that they escape attention. Each of them deflates grandiose schemes and conceptual dichotomies and replaces them by simple modifications in the way in which groups of people argue with one another using paper, signs, prints and diagrams. Despite their different methods, fields, and goals, this strategy of deflation links a range of very different studies and endows them with a style which is both ironic and refreshing.[2]

Like these scholars, I was struck, in a study of a biology laboratory, by the way in which many aspects of laboratory practice could be ordered by looking not at the scientists' brains (I was forbidden access!), at the cognitive structures (nothing special), nor at the paradigms (the same for thirty years), but at the transformation of rats and chemicals into paper (Latour and Woolgar, 1979/1986). Focusing on the literature, and the way in which anything and everything was transformed into inscriptions, was not my bias, as I first thought, but was for what the laboratory was made. Instruments, for instance, were of various types, ages, and degrees of sophistication. Some were pieces of furniture; others filled large rooms, employed many technicians, and took many weeks to run. But their end result, no matter the field, was always a small window through which one could read a very few signs from a rather poor repertoire (diagrams, blots, bands, columns). All these inscriptions, as I called them, were combinable, superimposable and could, with only a minimum of cleaning up, be integrated as figures in the text of the articles people were writing. Many of the intellectual feats I was asked to admire could be rephrased as soon as this activity of paper writing and inscription became the focus for analysis. Instead of jumping to explanations involving high theories or differences in logic, I could cling to the level of simple craftsmanship as firmly as Goody. The domestication or disciplining of the mind was still going on with instruments similar to those to which Goody refers. When these resources were lacking, the selfsame scientists stuttered, hesitated, and talked nonsense, and displayed every kind of political or cultural bias. Although their minds, their scientific methods, their paradigms, their world-views, and their cultures were still present, their conversation could not keep them in their proper place. However, inscriptions or the practice of inscribing could.

The Great Divide can be broken down into many small, unexpected and practical sets of skills to produce images, and to read and write about them. But there is a major drawback with this strategy of deflation. Its results seem both obvious—close to being literally a cliché—and too weak to account for the vast consequences of science and technology that cannot, we agreed above, be denied. Of course, everyone might happily agree that writing, printing, and visualizing are important *asides* of the scientific revolution or of the psychogenesis of scientific thought. They might be necessary but they certainly cannot be sufficient causes. Certainly not. The deflating strategy may rid us of one mystical Great Divide, but it will, it seems, lead us into a worse kind of mysticism if the researcher who deals with prints and images has to believe in the power of signs and symbols isolated from anything else.

This is a strong objection. We must admit that when talking of images and print it is easy to shift from the most powerful explanation to one that is trivial and reveals only marginal aspects of the phenomena for which we want to account. Diagrams, lists, formulae, archives, engineering drawings, files, equations, dictionaries, collections, and so on, depending on the way they are put into focus, may explain almost everything or almost nothing. It is all too easy to throw a set of clichés together extending Havelock's argument about the Greek alphabet (1980), or Walter Ong's rendering of the Ramist method (1971), all the way to computer culture, passing through the Chinese obsession with ideograms, double-entry bookkeeping—and without forgetting the Bible. Everyone agrees that print, images, and writing are everywhere present, but how much explanatory burden can they carry? How many cognitive abilities may be, not only facilitated, but thoroughly explained by them? When wading through this literature, I have a sinking feeling that we are alternately on firm new ground and bogged down in an old marsh. I want to find a way to hold the focus firmly so that we know what to expect from our deflating strategy.

To get this focus, first we must consider in which situations we might expect changes in the writing and imaging procedures to make any difference at all in the way we argue, prove, and believe. Without this preliminary step, inscriptions will, depending on the context, be granted either too much or too little weight.

Unlike Leroi-Gourhan (1964) we do not wish to consider all the history on writing and visual aids starting with primitive man and ending up with modern computers. From now on, we will be interested only in a few specific inventions in writing and imaging. To define this specificity we have to look more closely at the construction of harder facts.[3]

Who will win in an agonistic encounter between two authors and between them and all the others they need to build up a statement *S*? Answer: the one able to *muster on the spot the largest number of well aligned and faithful allies*. This definition of victory is common to war, politics, law, and, I shall now show, to science and technology. My contention is that writing and imaging cannot by themselves explain the changes in our scientific societies, except insofar as *they help to make this agonistic situation more favorable.* Thus it is not all the anthropology of writing, nor all the history of visualization, that interests us in this context. Rather, we should concentrate on those aspects that help in the mustering, the presentation, the increase, the effective alignment, or ensuring the fidelity of new allies. We need, in other words, to look at the way in which someone convinces someone else to take up a statement, to pass it along, to make it more of a fact, and to recognize the first author's ownership and originality. This is what I call "holding the focus steady" on visualization and cognition. If we remain at the level of the visual aspects only, we fall back into a series of weak clichés or are led into all sorts of fascinating problems of scholarship far away from our problem; but, on the other hand, if we concentrate on the agonistic situation alone, the principle of any victory, any solidity in science and technology escapes us forever. We have to hold the two eyepieces together so that we turn it into a real *bi*nocular; it takes time to focus, but the spectacle, I hope, is worth the waiting.

One example will illustrate what I mean. La Pérouse travels through the Pacific for Louis XVI with the explicit mission of bringing *back* a better map. One day, landing on what he calls Sakhalin, he meets with Chinese and tries to learn from them whether Sakhalin is an island or a peninsula. To his great surprise the Chinese understand geography quite well. An older man stands up and draws a map of his island on the sand with the scale and the details needed by La Pérouse. Another, who is younger, sees that the rising tide will soon erase the map and picks up one of La Pérouse's notebooks to draw the map again with a pencil. . .

What are the differences between the savage geography and the civilized one? There is no need to bring a prescientific mind into the picture, nor any distinction between the closed and open predicaments (Horton, 1977), nor primary and secondary theories (Horton, 1982), nor divisions between implicit and explicit, or concrete and abstract geography. The Chinese are quite able to think in terms of a map but also to talk about navigation on an equal footing with La Pérouse. Strictly speaking, the ability to draw and to visualize does not really make a difference either, since they all draw maps more or less based on the same principle of projection, first on sand, then on paper. So perhaps there is no difference after all and, geographies being equal, relativism is right? This, however, cannot be, because La Pérouse does something that is going to create an enormous difference between the Chinese and the European. What is, for the former, a drawing of no importance that the tide may erase, is for the latter the *single object* of his mission. What should be brought into the picture is how the picture is brought back. The Chinese does not have to keep track, since he can generate many maps at will, being born on this island and fated to die on it. La Pérouse is not going to stay for more than a night; he is not born here and will die far away. What is he doing, then? He is passing through all these places, in order to take something *back* to Versailles where many people expect his map to determine who was right and wrong about whether Sakhalin was an island, who will own this and that part of the world, and along which routes the next ships should sail. Without this peculiar trajectory, La Pérouse's exclusive interest in traces and inscriptions will be impossible to understand—this is the first aspect; but without dozens of innovations in inscription, in projection, in writing, archiving, and computing, his displacement through the Pacific would be totally wasted—and this is the second aspect, as crucial as the first. We have to hold the two together. Commercial interests, capitalist spirit, imperialism, thirst for knowledge, are empty terms as long as one does not take into account Mercator's projection, marine clocks and their makers, copper engraving of maps, rutters, the keeping of "log books," and the many printed editions of Cook's voyages that La Pérouse carries with him. This is where the deflating strategy I outlined above is so powerful. But, on the other hand, no innovation in the way longitude and latitudes are calculated, clocks are built, log books are compiled, copper plates are printed, would make any difference whatsoever if they did not help to muster, align, and win over new and unexpected allies, far away in Versailles. The practices I am interested in would be pointless if they did not bear on certain controversies and force dissenters into believing new facts and behaving in new ways. This is where an exclusive interest in visualization and writing falls short, and can even be counterproductive. To maintain only the second line of argument would offer a mystical view of the powers provided by semiotic material—as did Derrida (1967); to maintain only the first would be to offer an idealist explanation (even if clad in materialist clothes).

The aim of this paper is to pursue the two lines of argument at once. To say it in yet other words, we do not find all explanations in terms of inscription equally convincing, but only those that help us to understand how the mobilization and mustering of new resources is achieved. We do not find all explanations in terms of social groups, interests or economic trends, equally convincing but only those that offer a specific mechanism to sum up "groups," "interests," "money," and "trends": mechanisms which we believe, depend upon the manipulation of paper, print, images, and so on. La Pérouse shows us the way since without new types of inscriptions nothing usable would have come back to Versailles from his long, costly, and fateful voyage; but without this strange mission that required him to go away and to come back so that others in France might be convinced, no modification in inscription would have made a bit of difference.

The essential characteristics of inscriptions cannot be defined in terms of visualization, print, and writing. In other words, it is not *perception* which is at stake in this problem of visualization and cognition. New inscriptions, and new ways of perceiving them, are the results of something deeper. If you wish to go out of *your* way and come back heavily equipped so as to force others to go out of *their* ways, the main problem to solve is that of *mobilization*. You have to go and to come back *with* the "things" if your moves are not to be wasted. But the "things" have to be able to withstand the return trip without withering away. Further requirements: the "things" you gathered and displaced have to be presentable all at once to those you want to convince and who did not go there. In sum, you have to invent objects which have the properties of being *mobile* but also *immutable*, *presentable*, *readable* and *combinable* with one another.

2. On immutable mobiles

It seems to me that most scholars who have worked on the relations between inscription procedures and cognition, have, in fact, in their various ways, been writing about the history of these immutable mobiles.

2.1 Optical consistency

The first example I will review is one of the most striking since Ivins wrote about it years ago and saw it all in a few seminal pages. The rationalization that took place during the so-called "scientific revolution" is not of the mind, of the eye, of philosophy, but of the *sight*. Why is perspective such an important invention? "Because of its logical recognition of internal invariances through all the transformations produced by changes in spatial location" (Ivins, 1973:9). In linear perspective, no matter from what distance and angle an object is seen, it is always possible to transfer it—to translate it—and to obtain the same object at a different size as seen from another position. In the course of this translation, its internal properties have not been modified. This immutability of the displaced figure allows Ivins to make a second crucial point: since the picture moves without distortion it is possible to establish, in the linear perspective framework, what he calls a "two-way" relationship between object and figure. Ivins shows us how perspective allows movement through space with, so to speak, a return ticket. You can see a church in Rome, and carry it with you in London in such a way as to reconstruct it in London, or you can go back to Rome and amend the picture. With perspective exactly as with La Pérouse's map—and for the same reasons—a new set of movements are made possible: you can go out of your way and come back with all the places you passed; these are all written in the same homogeneous language (longitude and latitude, geometry) that allows you to change scale, to make them presentable, and to combine them at will.[4]

Perspective, for Ivins, is an essential determinant of science and technology because it creates "optical consistency," or, in simpler terms, a regular avenue through space. Without it "either the exterior relations of objects such as their forms for visual awareness, change with their shifts in locations, or else their interior relations do" (1973:9). The shift from the other senses to vision is a consequence of the agonistic situation. You present absent things. No one can smell or hear or touch Sakhalin island, but you can look at the map and determine at which bearing you will see the land when you send the next fleet. The speakers are talking to one another, feeling, hearing and touching each other, *but* they are now talking *with* many absent things presented all at once. This presence/absence is possible through the two-way connection established by these many contrivances—perspective, projection, map, logbook, etc.—that allow translation without corruption.

There is another advantage of linear perspective to which he and Edgerton attract our attention (1976). This unexpected advantage is revealed as soon as religious or mythological themes and utopias are drawn with the same perspective as that which is used for rendering nature (Edgerton, 1980:189).

> In the West, even if the subject of the printed text were unscientific, the printed picture always presented a rational image based on the universal laws of geometry. In this sense the Scientific Revolution probably owes more to Albrecht Dürer than to Leonardo da Vinci. (1980:190)

Fiction—even the wildest or the most sacred—and things of nature—even the lowliest—have a meeting ground, *a common place*, because they all benefit from the same "optical consistency."[5] Not only can you displace cities, landscapes, or natives and go back and forth to and from them along avenues through space, but you can also reach saints, gods, heavens, palaces, or dreams with the same two-way avenues and look at them through the same "windowpane" on the same two-dimensional surface. The two ways become a four-lane freeway! Impossible palaces can be drawn realistically, but it is also possible to draw possible objects as if they were utopian ones. For instance, as Edgerton shows, when he comments on Agricola's prints, real objects can be drawn in separated pieces, or in exploded views, or added to the same sheet of paper at different scales, angles and perspectives. It does not matter since the "optical consistency" allows all the pieces to mix with one another.

> Oddly enough, linear perspective and chiaroscuro, which supply geometric stability to pictures, also allow the viewer a momentary suspension of his dependence on the law of gravity. With a little practice, the viewer can imagine solid volumes floating freely in space as detached components of a device. (Edgerton, 1980:193)

As Ferguson says, the "mind" has at last "an eye".

At this stage, on paper, hybrids can be created that mix drawings from many sources. Perspective is not interesting because it provides realistic pictures; on the other hand, it is interesting because it creates complete hybrids: nature seen as fiction, and fiction seen as nature, with all the elements made so homogeneous in space that it is now possible to reshuffle them like a pack of cards. Commenting on the painting "St. Jérome in his study," Edgerton says:

> Antonello's St. Jérome is the perfect paradigm of a new consciousness of the physical world attained by Western European intellectuals by the late fifteenth century. This consciousness was showed especially by artists such as Leonardo da Vinci, Francesco di Giorgio Martini, Albrecht Dürer, Hans Holbein and more, all of whom . . . had even developed a sophisticated grammar and syntax for quantifying natural phenomena in pictures. In their hands, picture making was becoming a pictorial language that, with practice, could communicate more information, more quickly and by (sic) a potentially wider audience than any verbal language in human history. (1980:189)

Perspective illustrates the double line of argument I presented in the previous section. Innovations in graphism are crucial but only insofar as they allow new two-way relations to be established with objects (from nature or from fiction) and only insofar as they allow inscriptions either to become more mobile or to stay immutable through all their displacements.

2.2 *Visual culture*

Still more striking than the Italian perspective described by Ivins and Edgerton, is the Dutch "distance point" method for drawing pictures, as it has been beautifully explained by Svetlana Alpers (1983). The Dutch, she tells us, do not paint grandiose historical scenes as observed by someone through a carefully framed windowpane. They use the very surface of their paintings (taken as the equivalent of a retina) to let the world be painted straight on it. When images are captured in this way there is no privileged site for the onlooker any more. The tricks of the camera obscura transform large-scale three-dimensional objects into a small two-dimensional surface around which the onlooker may turn at will.[6]

The main interest of Alpers' book for our purpose is the way she shows a "visual culture" changing over time. She does not focus on the inscriptions or the pictures but on the simultaneous transformation of science, art, theory of vision, organization of crafts and economic powers. People often talk of "worldviews," but this powerful expression is taken metaphorically. Alpers provides this old expression with its material meaning: how a culture *sees* the *world,* and makes it visible. A new visual culture redefines both what it is to see, and what there is to see. A citation of Comenius aptly summarizes a new obsession for making objects visible anew:

> We will now speak of the mode in which objects must be presented to the senses, if the impression is to be distinct. This can be readily understood if we consider the process of actual vision. If the object is to be clearly seen it is necessary: (1) that it be placed before the eyes; (2) not far off, but at a reasonable distance; (3) not on one side, but straight before the eyes; and (4) so that the front of the objects be not turned away from, but directed towards, the observer; (5) that the eyes first take in the object as a whole; (6) and then proceed to distinguish the parts; (7) inspecting these in order from the beginning to the end; (8) that attention be paid to each and every part; (9) until they are all grasped by means of their essential attributes. If these requisites be properly observed, vision takes place successfully; but if one be neglected its success is only partial. (cited in Alpers, 1983:95)

This new obsession for defining the act of seeing is to be found both in the science of the period and in modern laboratories. Comenius' advice is similar to both that of Boyle when he disciplined the witnesses of his air-pump experiment (Shapin, 1984) and that of the neurologists studied by Lynch when they "disciplined" their brain cells (Lynch, 1985a). People before science and outside laboratories certainly use their eyes, but not in this way. They look at the spectacle of the world, but not at this new type of image designed to transport the objects of the world, to accumulate them in Holland, to label them with captions and legends, to combine them at will. Alpers makes understandable what Foucault (1966) only suggested: how the same eyes suddenly began to look at "representations." The "panopticon" she describes is a *fait social total* that redefines all aspects of the culture. More importantly, Alpers does not explain a new vision by bringing in "social interests" or the "economic infrastructure." The new precise scenography that results in a worldview defines at once what is science, what is art, and what it is to have a world economy. To use my terms, a little lowland country becomes powerful by making a few crucial inventions which allow people to accelerate the mobility and to enhance the immutability of inscriptions: the world is thus gathered up in this tiny country.

Alpers' description of Dutch visual culture reaches the same result as Edgerton's study of technical drawings: a new meeting place is designed for fact and fiction, words and images. The map itself is such a result, but the more so when it is used to inscribe ethnographic inventories (end of her chapter IV) or captions (chapter V), skylines of cities and so on. The main quality of the new space is not to be "objective," as a naïve definition of realism often claims, but rather to have optical consistency. This consistency entails the *art of describing* everything and the possibility of going from one type of visual trace to another. Thus, we are not surprised that letters, mirrors, lenses, painted words, perspectives, inventories, illustrated children's books, microscopes, and telescopes come together in this visual culture. All innovations are selected "to secretly see and without suspicion what is done far off in other places" (cited in Alpers, 1983:201).

2.3 A new way of accumulating time and space

Another example will demonstrate that inscriptions are not interesting per se but only because they increase either the mobility or the immutability of traces. The invention of print and its effects on science and technology is a cliché of historians. But no one has renewed this Renaissance argument as completely as Elizabeth Eisenstein (1979). Why? Because she considers the printing press to be a mobilization device, or, more exactly, a device that makes both mobilization and immutability possible at the same time. Eisenstein does not look for one cause of the scientific revolution, but for a secondary cause that would put all the efficient causes in relation with one another. The printing press is obviously a powerful cause of that sort. Immutability is ensured by the process of printing many identical copies; mobility by the number of copies, the paper, and the movable type. The links between different places in time and space are completely modified by this fantastic acceleration of immutable mobiles which circulate everywhere in all directions in Europe. As Ivins has shown, perspective plus the printing press plus aqua forte is the really important combination since books can now carry with them the realistic images of what they talk about. For the first time, a location can accumulate other places far away in space and time, and present them synoptically to the eye; better still, this synoptic presentation, once reworked, amended, or disrupted, can be spread with no modification to other places and made available at other times.

After discussing historians who propose many contradictory influences to explain the take-off of astronomy, Eisenstein writes:

> Whether the sixteenth-century astronomer confronted materials derived from the fourth century B.C. or freshly composed in the fourteenth century A.D., or whether he was more receptive to scholastic or humanist currents of thoughts, seems of less significance in this particular connection than the fact that all manners of diverse materials were being seen in the course of one life time by one pair of eyes. For Copernicus as for Tycho, the result was heightened awareness and dissatisfaction with discrepancies in the inherent data. (1979:602)

Constantly, the author shifts attention with devastating irony from the mind to the surface of the mobilized resources:

> "To discover the truth of a proposition in Euclid," wrote John Locke, "there is little need or use of revelation, God having furnished us with a natural and surer means to arrive at knowledge of them." In the eleventh century, however, God had not furnished Western scholars with a natural and sure means of grasping a Euclidean theorem. Instead the most learned men in Christendom engaged in a fruitless search to discover what Euclid meant when referring to interior angles. (1979:649)

For Eisenstein, every grand question about the Reformation, the Scientific Revolution, and the new Capitalist economy can be recast by looking at what the publisher and the printing press make possible. The reason why this old explanation takes on new life in her treatment is that Eisenstein not only focuses on graphism, but also on changes in the graphism that are linked to the mobilization process. For instance, she explains (p. 508 ff. following Ivins, 1953) the puzzling phenomenon of a lag time between the introduction of the printing press and the beginning of exact realistic pictures. At first, the press is used simply to reproduce herbaries, anatomical plates, maps, and cosmologies that are centuries old and that will be deemed inaccurate much later. If we were looking only at the semiotic level this phenomenon would seem puzzling, but once we consider the deeper structure this is easily explained. The displacement of many immutable mobiles comes first; the old texts are spread everywhere and can be gathered more cheaply in one place. But then the contradiction between them at last becomes *visible* in the most literal sense. The many places where these texts are synoptically assembled offer many counterexamples (different flowers, different organs with different names, different shapes for the coastline, the various rates of different currencies, different laws). These counterexamples can be added to the old texts and, in turn, are spread without modification to all the other settings where this process of comparison may be resumed. In other words, errors are accurately reproduced and spread with no changes. But corrections are also reproduced fast, cheaply and with no further changes. So, at the end, the accuracy *shifts from the medium to the message*, from the printed book to the context with which it establishes a two-way connection. A new interest in "Truth" does not come from a new vision, but from the same old vision applying itself to new visible objects that mobilize space and time differently.[7]

The effect of Eisenstein's argument is to transform mentalist explanations into the history of immutable mobiles. Again and again she shows that before the advent of print every possible intellectual feat had been achieved—organized scepticism, scientific method, refutation, data collection, theory making—everything had been tried, and in all disciplines: geography, cosmology, medicine, dynamics, politics, economics, and so on. But each achievement stayed local and temporary just because there was no way to move their results elsewhere and to bring in those of others without new corruptions or errors being introduced. For instance, each carefully amended version of an old author was, after a few copies, again adulterated. No irreversible gains could be made, and so no large-scale long-term capitalization was possible. The printing press does not add anything to the mind, to the scientific method, to the brain. It simply conserves and spreads everything no matter how wrong, strange, or wild. It makes everything mobile but this mobility is not offset by adulteration. The new scientists, the new clerics, the new merchants, and the new princes, described by Eisenstein, are no different from the old ones, but they now look at new material that keeps track of numerous places and times. No matter how inaccurate these traces might be at first, they will all become accurate just *as a consequence* of more mobilization and more immutability. A mechanism is invented to irreversibly capture accuracy. Print plays the same role as Maxwell's demon. No new theory, worldview, or spirit is necessary to explain capitalism, the reformation, and science: they are the result of a new step in the long history of immutable mobiles.

Taking up Ivins' argument, both Mukerji (1983) and Eisenstein focus again on the *illustrated* book. For these authors, MacLuhan's revolution had already happened as soon as images were printed. Engineering, botany, architecture, mathematics, none of these sciences can describe what they talk about with texts alone; they need to show the things. But this showing, so essential to convince, was utterly impossible before the invention of "graven images." A text could be copied with only some adulteration, but not so a diagram, an anatomical plate, or a map. The effect on the construction of facts is sizable if a writer is able to provide a reader with a text that presents a large number of the things it is talking about in one place. If you suppose that all the readers and all the writers are doing the same, a new world will emerge from the old one without any additional cause. Why? Simply because the dissenter will have to do the same thing as his opponent. In order to "doubt back," so to speak, he will have to write another book, have it printed, and mobilize with copper plates the counterexamples he wants to oppose. The cost of disagreeing will increase.[8]

Positive feedback will get under way as soon as one is able to muster a large number of mobile, readable, visible resources at one spot to support a point. After Tycho Brahe's achievement (Eisenstein, 1979) the dissenter either has to quit and accept what cosmologists say as a hard fact, or to produce counterproofs by persuading his prince to invest a comparable amount of money in observatories. In this, the "proof race" is similar to the arms race because the feedback mechanism is the same. Once one competitor starts building up harder facts, the others have to do the same or else submit.

This slight recasting of Eisenstein's argument in terms of immutable mobiles may allow us to overcome a difficulty in her argument. Although she stresses the importance of publishers' strategies, she does not account for the technical innovations themselves. The printing press barges into her account like the exogeneous factors of many historians when they talk about technical innovations. She puts the semiotic aspect of print and the mobilization it allows into excellent focus, but the technical necessities for inventing the press are far from obvious. If we consider the agonistic situation I use as reference point, the pressure that favors something like the printing press is clearer. *Anything* that will accelerate the mobility of the traces that a location may obtain about another place, or *anything* that will allow these traces to move without transformation from one place to another, will be favored: geometry, projection, perspective, bookkeeping, paper making, aqua forte, coinage, new ships (Law, 1986). The privilege of the printing press comes from its ability to help many innovations to act at once, but it is only one innovation among the many that help to answer this simplest of all questions: how to dominate on a large scale? This recasting is useful since it helps us to see that the same mechanism, the effects of which are described by Eisenstein, *is still at work* today, on an ever-increasing scale at the frontiers of science and technology. A few days in a laboratory reveal that the same trends that made the printing press so necessary, still act to produce new data bases, new space telescopes, new chromatographies, new equations, new scanners, new questionnaires, etc. The mind is still being domesticated.

3. On inscriptions

What is so important in the images and in the inscriptions scientists and engineers are busy obtaining, drawing, inspecting, calculating, and discussing? It is, first of all, the unique advantage they give in the rhetorical or polemical situation. "You doubt what I say? I'll show you." And, without moving more than a few inches, I unfold in front of your eyes figures, diagrams, plates, texts, silhouettes, and then and there present things that are far away and with which some sort of two-way connection has now been established. I do not think the importance of this simple mechanism can be overestimated. Eisenstein has shown it for the past of science, but ethnography of present laboratories shows the same mechanism (Lynch, 1985a, 1985b; Star, 1983; Law, 1985). We are so used to this world of print and images, that we can hardly think of what it is to know something without indexes, bibliographies, dictionaries, papers with references, tables, columns, photographs, peaks, spots, bands.[9]

One simple way to make the importance of inscriptions clearer is to consider how little we are able to convince when deprived of these graphisms through which mobility and immutability are increased. As Dagognet has shown in two excellent books, no scientific discipline exists without first inventing a visual and written language which allows it to break with its confusing past (1969, 1973). The manipulation of substances in gallipots and alambics becomes chemistry only when all the substances can be written in a homogeneous language where everything is simultaneously presented to the eye. The writing of words inside a classification are not enough. Chemistry becomes powerful only when a visual vocabulary is invented that replaces the manipulations by calculation of formulas. Chemical structure can be drawn, composed, broken apart on paper, like music or arithmetic, all the way to Mendeleiev's table: "for those who know to observe and read the final periodic table, the properties of the element and that of their various combinations unfold completely and directly from their positions in the table" (1969:213). After having carefully analyzed the many innovations in chemical writing and drawings, he adds this little sentence so close to Goody's outlook:

> It might seem that we consider trivial details—a slight modification in the plane used to write a chlorine—but, paradoxically, these little details trigger the forces of the modern world. (1969:p. 199)

Michel Foucault, in his well-known study of clinical medicine, has shown the same transformation from small-scale practice to a large-scale manipulation of records (1963). The same medical mind will generate totally different knowledge if applied to the bellies, fevers, throats, and skins of a few successive patients, or if applied to well-kept records of hundreds of written bellies, fevers, throats, and skins, all coded in the same way and all synoptically present. Medicine does not become scientific in the mind, or in the eye of its practitioners, but in the application of old eyes and old minds to new fact sheets inside new institutions, the hospital. But it is in *Discipline and Punish* (1975) that Foucault's demonstration is closest to the study of inscriptions. The main purpose of the book is to illustrate the shift from a power which is seen by invisible onlookers, to a new invisible power that sees everything about everyone. The main advantage of Foucault's analysis is not to focus only on files, accounting books, time tables, and drill, but also on the sort of institutions in which these inscriptions end up being so essential.[10] The main innovation is that of a "panopticon" which allows penology, pedagogy, psychiatry, and clinical medicine to emerge as full-fledged sciences from their carefully kept files. The panopticon is another way of obtaining the "optical consistency" necessary for power on a large scale.

In a famous sentence, Kant asserts that "we shall be rendering a service to reason should we succeed in discovering the path upon which it can securely travel." The "sure path of a science," however, is, inevitably, in the construction of well-kept files in institutions that want to mobilize a larger number of resources on a larger scale.

"Optical consistency" is obtained in geology, as Rudwick has shown (1976), by inventing a new visual language. Without it, the layers of the earth stay hidden and no matter how many travellers and diggers move around there is no way to sum up their travels, visions, and claims. The Copernican revolution, dear to Kant's heart, is an idealist rendering of a very simple mechanism: if we cannot go to the earth, let the earth come to us, or, more accurately, let us all go to many places on the earth, and come back with the same but different homogenous pictures, that can be gathered, compared, superimposed, and redrawn in a few places, together with the carefully labelled specimens of rocks and fossils.

In a suggestive book, Fourquet (1980) has illustrated the same inscription gathering for INSEE, the French institution that provides most economic statistics. It is of course impossible to talk about the economy of a nation by looking at "it." The "it" is plainly invisible, as long as cohorts of enquirers and inspectors have not filled in long questionnaires, as long as the answers have not been punched onto cards, treated by computers, analyzed in this gigantic laboratory. Only at the end can the economy be made visible inside piles of charts and lists. Even this is still too confusing, so that redrawing and extracting is necessary to provide a few neat diagrams that show the Gross National Product or the Balance of Payments. The panopticon thus achieved is similar in structure to a gigantic scientific instrument transforming the invisible world of exchanges into "the economy." This is why, at the beginning, I rejected the materialist explanation that uses "infrastructures" or "markets" or "consumer needs" to account for science and technology. The visual construction of something like a "market" or an "economy" is what begs explanation, and this end-product cannot be used to account for science.

In another suggestive book Fabian tries to account for anthropology by looking at its craftsmanship of visualization (1983). The main difference between us and the savages, he argues, is not in the culture, in the mind, or in the brain, but in the way *we* visualize *them*. An asymmetry is created because we create a space and a time in which we place the other cultures, but they do not do the same. For instance, we map their land, but they have no maps either of their land or of ours; we list their past, but they do not; we build written calendars, but they do not. Fabian's argument, related to Goody's and also to Bourdieu's critique of ethnography (1972), is that once this first violence has been committed, no matter what we do, we will not understand the savages any more. Fabian however, sees this mobilization of all savages in a few lands through collection, mapping, list making, archives, linguistics, etc., as something evil. With candor, he wishes to find another way to "know" the savages. But "knowing" is not a disinterested cognitive activity; harder facts about the other cultures have been produced in our societies, in exactly the same way as other facts about ballistics, taxonomy or surgery. One place gathers in all the others and presents them synoptically to the dissenter so as to modify the outcome of an agonistic encounter. To make a large number of competitors and compatriots depart from their usual ways, many ethnographers both had to go further and longer out of *their* usual ways, and then come *back*. The constraints imposed by convincing people, going out and coming back, are such that this can be achieved only if everything about the savage life is transformed into immutable mobiles that are easily readable and presentable. In spite of his wishes, Fabian cannot do better. Otherwise, he would either have to give up "knowing" or give up making hard facts (Latour, 1987).

There is no detectable difference between natural and social science, as far as the obsession for graphism is concerned. If scientists were looking at nature, at economies, at stars, at organs, they would not *see* anything. This "evidence," so to speak, is used as a classic rebuttal to naïve versions of empiricism (Arnheim, 1969). Scientists start seeing something once they stop looking at nature and look exclusively and obsessively at prints and flat inscriptions.[11] In the debates around perception, what is always forgotten is this simple drift from watching confusing three-dimensional objects, to inspecting two-dimensional images which have been *made less confusing*. Lynch, like all laboratory observers, has been struck by the extraordinary obsession of scientists with papers, prints, diagrams, archives, abstracts and curves on graph paper. No matter what they talk about, they start talking with some degree of confidence and being believed by colleagues, only once they point at simple geometrized two-dimensional shapes. The "objects" are discarded or often absent from laboratories. Bleeding and screaming rats are quickly dispatched. What is extracted from them is a tiny set of figures. This extraction, like the few longitudes and latitudes extracted from the Chinese by La Pérouse, is *all that counts*. Nothing can be said about the rats, but a great deal can be said about the figures (Latour and Woolgar, 1979). Knorr (1981) and Star (1983) have also shown the simplification procedures at work, as if the images were never simple enough for the controversy to be settled quickly. Every time there is a dispute, great pains are taken to find, or sometimes to invent, a new instrument of visualization, which will enhance the image, accelerate the readings, and, as

Lynch has shown, conspire with the visual characteristics of the things that lend themselves to diagrams on paper (coast lines, stars which are like points, well-aligned cells, etc.).

Again, the precise focus should be carefully set, because it is not the inscription by itself that should carry the burden of explaining the power of science; it is the inscription *as the fine edge* and *the final stage* of a whole process of mobilization, that modifies the scale of the rhetoric. Without the displacement, the inscription is worthless; without the inscription the displacement is wasted. This is why mobilization is not restricted to paper but paper always appears at the end when the scale of this mobilization is to be increased. Collections of rocks, stuffed animals, samples, fossils, artifacts, gene banks, are the first to be moved around (Star and Griessemer, 1989). What counts is the arraying and mustering of resources (biographies of naturalists, for instance, are replete with anecdotes about crates, archives and specimens), but this arraying is never simple enough. Collections are essential but only while the archives are well-kept, the labels are in place, and the specimens do not decay. Even this is not enough, since a museum collection is still too much for one "mind" to handle. So the collection will be drawn, written, recorded, and this process will take place as long as more combinable geometrized forms have not been obtained from the specimens (continuing the process through which the specimens had been extracted from their contexts).

So, the phenomenon we are tackling is *not* inscription per se, but the *cascade* of ever simplified inscriptions that allow harder facts to be produced at greater cost. For example, the description of human fossils which used to be through drawings, is now made by superimposing a number of mechanical diagrams on the drawings. The photographs of the skies, although they produce neat little spots, are still much too rich and confusing for a human eye to look at; so a computer and a laser eye have been invented to read the photographs, so that the astronomer never looks at the sky (too costly), nor even at the photographs (too confusing). The taxonomy of plants is all contained in a famous series of books at Kew Garden, but the manipulation of this book is as difficult as that of the old manuscripts since it exists in only one location; another computer is now being instructed to try to read the many different prints of this book and provide as many copied versions as possible of the taxonomic inventory.

Pinch (1985) shows a nice case of accumulation of such traces, each layer being deposited on the former one only when confidence about its meaning is stabilized. Do the astrophysicists "see" the neutrinos from the sun or any of the intermediary "blurs," "peaks," and "spots" that compose, by accumulation, the phenomenon to be seen? Again, we see that the mechanisms studied by Eisenstein for the printing press are still with us today at any of the frontiers of science. For instance, baboon ethology used to be a text in prose in which the narrator talked about animals; the narrator had to include in the text what he or she had seen first as pictures, and then a statistical rendering of the events; but with an increasing competition for the construction of harder facts, the articles now include more and more layers of graphic display, and the cascade of columns summarized by tables, diagrams, and equations is still unfolding. In molecular biology, chromatography was read, a few years ago, by bands of different shades of gray; the interpretation of these shades is now done by computer, and a text is eventually obtained straight out of the computer: "ATGCGTTCGC. . . ." Although more empirical studies should be made in many different fields, there seems to be a trend in these cascades. They always move on the direction of the greater merging of figures, numbers, and letters, which is greatly facilitated by their homogeneous treatment as binary units in and by computers.

This trend toward simpler and simpler inscriptions that mobilize larger and larger numbers of events in one spot cannot be understood if separated from the agonistic model that we use as our point of reference. It is as necessary as the race for digging trenches on the front in 1914. He who visualizes badly loses the encounter; his fact does not hold. Knorr has criticised this argument by taking an ethnomethodological standpoint (1981). She argues, and rightly so, that an image, a diagram, cannot convince anyone, both because there are always many interpretations possible, and, above all, because the diagram does not force the dissenter to look at it. She sees the interest in inscription devices as an exaggeration of the power of semiotics (and a French one at that!). But such a position misses the point of my argument. It is precisely because the dissenter can always escape and try out another interpretation, that so much energy and time is devoted by scientists to *corner* him and surround him with ever more dramatic visual effects. Although *in principle* any interpretation can be opposed to any text and image, *in practice* this is far from being the case; the cost of dissenting increases with each new collection, each new labelling, each new redrawing. This is especially true if the phenomena we are asked to believe are invisible to the naked eye; quasars, chromosomes, brain peptides, leptons, gross national products, classes, and coastlines are never seen but through the "clothed" eye of inscription devices. Thus, one *more* inscription, one more trick to enhance contrast, one simple device to decrease background, or one coloring procedure might be *enough,* all things being equal, *to swing the balance* of power and turn an incredible statement into a credible one that would then be passed along without further modification. The importance of this cascade of inscriptions may be ignored when studying events in daily life, but it cannot be overestimated when analyzing science and technology.

More exactly, it is possible to overestimate the inscription, but not the setting in which the cascade of ever more written and numbered inscriptions is produced. What we are really dealing with is the *staging* of a scenography in which attention is focused on one set of dramatized inscriptions. The setting works like a giant optical device that creates a new laboratory, a new type of vision, and a new phenomenon to look at. I showed one such setting which I called "Pasteur's theater of proofs" (Latour, 1988a). Pasteur works as much on the stage as on the scene and the plot. What counts at the end is a simple visual perception: dead unvaccinated sheep versus alive vaccinated sheep. The earlier we go back in the history of science, the more attention we see being paid to the setting and the less to inscriptions themselves. Boyle, for instance, in the fascinating account of his vacuum pump experiment described by Shapin (1984), had to invent not only the phenomenon, but the instrument to make it visible, the set-up in which the instrument was displayed, the written and printed accounts through which the silent reader could read "about" the experiment, the type of witnesses admitted onto the stage, and even the types of commentaries the potential witnesses were allowed to utter. "Seeing the vacuum" was possible only once all these witnesses had been disciplined.

The staging of such "optical devices" is the one Eisenstein describes: a few persons in the same room talk to one another and point out two-dimensional pictures; these pictures are all there is to see of the things about which they talk. Just because we are used to this setting, and breathe it like fresh air, does not mean that we should not describe all the little innovations that make it the most powerful device to achieve power. Tycho Brahe, in Oranienburg, had before his eyes for the first time in history all the predictions—that is literally the "pre*visions*"—of the planetary movements; at the *same* place, written in the *same* language or code, he can read *his* own observations. This is more than enough to account for Brahe's new "insight."

> It was not because he gazed at night skies instead of at old books that Tycho Brahe differed from star-gazers of the past. Nor do I think it was because he cared more for "stubborn facts" and precise measurement than had the Alexandrians or the Arabs. But he did have at his disposal, as few had before him, two separate sets of computations based on two different theories, compiled several centuries apart which he could compare with each other. (Eisenstein, 1979:624)

Historians say that he is the first to look at planetary motion, with a mind freed of the prejudices of the darker ages. No, says Eisenstein, he is the first *not* to look at the sky, but to look simultaneously to all the former predictions and his own, written down together in the same form.

> The Danish observer was not only the last of the great naked eye observers; he was also the first careful observer who took full advantage of the new powers of the press—powers which enabled astronomers to detect anomalies in old records, to pinpoint more precisely and register in catalogs the location of each star, to enlist collaborators in many regions, fix each fresh observation in permanent form and make necessary corrections in successive editions. (1979:625)

The discrepancies proliferate, not by looking at the sky, but by carefully superimposing columns of angles and azimuths. No contradiction or counterpredictions could ever have been visible. Contradiction, as Goody says, is neither a property of the mind, nor of the scientific method, but is a property of reading letters and signs inside new settings that focus attention on inscriptions alone.

The same mechanism is visible, to draw an example from a different time and place, in Roger Guillemin's vision of endorphin, a brain peptide. The brain is as obscure and as messy as the Renaissance sky. Even the many first-level purifications of brain extracts provide a "soup" of substances. The whole research strategy is to get peaks that are clearly readable out of a confused background. Each of the samples which provides a neater peak is in turn purified until there is only one peak on the little window of a high pressure liquid chromatograph. Then the substance is injected in minute quantities into guinea-pig gut. The contractions of the gut are hooked up, through electronic hardware, to a physiograph. What is there at hand to see the object "endorphine"? The superimposition of the first peak with the slope in the physiograph starts to produce an object whose limits are the visual inscriptions produced in the lab. The object is a real object no more and no less than any other, since many such visual layers can be produced. Its resistance as a real fact depends only on the number of such visual layers that Guillemin's lab can mobilize all at once in one spot, in front of the dissenter. For each "objection" there is an inscription that blocks the dissent; soon, the dissenter is forced to quit the game or to come back later with the other and better visual displays. Objectivity is slowly erected inside the laboratory walls by mobilizing more faithful allies.

4. Capitalizing inscriptions to mobilize allies

Can we summarize why it is so important for Brahe, Boyle, Pasteur, or Guillemin to work on two-dimensional inscriptions instead of the sky, the air, health, or the brain? What can they do with the first that you cannot do with the second? Let me list a few of the advantages of "paperwork."

1. Inscriptions are *mobile,* as I indicated for La Pérouse's case. Chinese, planets, microbes—none of these can move; however, maps, photographic plates, and Petri dishes can.

2. They are *immutable* when they move, or at least everything is done to obtain this result: specimens are chloroformed, microbe colonies are stuck into gelatin, even exploding stars are recorded on graph paper in each phase of their explosion.

3. They are made *flat*. There is nothing you can *dominate* as easily as a flat surface of a few square meters; there is nothing hidden or convoluted, no shadows, no "double entendre." In politics as in science, when someone is said to "master" a question or to "dominate" a subject, you should normally look for the flat surface that enables mastery (a map, a list, a file, a census, the wall of a gallery, a card-index, a repertory) and you will find it.

4. The *scale* of the inscriptions may be *modified* at will, without any change in their internal proportions. Observers never insist on this simple fact: no matter what the (reconstructed) size of the phenomena, they all end up being studied only when they reach the same average size. Billions of galaxies are never bigger, when they are counted, than nanometer-sized chromosomes; international trade is never much bigger than mesons; scale models of oil refineries end up having the same dimensions as plastic models of atoms. Confusion resumes outside a few square meters. This trivial change of scale seems innocuous enough, but it is the cause of most of the "superiority" of scientists and engineers: no one else deals only with phenomena that can be dominated with the eyes and held by hands, no matter when and where they come from or what their original size.

5. They can be *reproduced* and spread at little cost, so that all the instants of time and all the places in space can be gathered in another time and place. This is "Eisenstein's effect."

6. Since these inscriptions are mobile, flat, reproducible, still, and of varying scales, they can be reshuffled and *recombined*. Most of what we impute to connections in the mind may be explained by this reshuffling of inscriptions that all have the same "optical consistency." The same is true of what we call "metaphor" (see Latour and Woolgar, 1979, chap. 4; Goody, 1977; Hughes, 1979; Ong, 1982).

7. One aspect of these recombinations is that it is possible to *superimpose* several images of totally different origins and scales. To link geology and economics seems an impossible task, but to superimpose a geological map with the printout of the commodity market at the New York Stock Exchange requires good documen-

tation and takes a few inches. Most of what we call "structure," "pattern," "theory," and "abstraction" are consequences of these superimpositions (Bertin, 1973). "Thinking is hand-work," as Heidegger said, but what is in the hands are inscriptions. Levi-Strauss' theories of savages are an artifact of card indexing at the College de France, exactly as Ramist's method is, for Ong, an artifact of the prints accumulated at the Sorbonne; or modern taxonomy a result of the bookkeeping undertaken, among other places, at Kew Gardens.

8. But one of the most important advantages is that the inscription can, after only little cleaning up, be *made part of a written text.* I have considered elsewhere at length this common ground in which inscriptions coming from instruments merge with already published texts and with new texts in draft. This characteristic of scientific texts has been demonstrated by Ivins and Eisenstein for the past. A present-day laboratory may still be defined as the unique place where a text is made to comment on things which are all present in it. Because the commentary, earlier texts (through citations and references), and "things" have the same optical consistency and the same semiotic homogeneity, an extraordinary degree of certainty is achieved by writing and reading these articles (Latour and Bastide, 1985; Lynch, 1985a; Law, 1983). The text is not simply "illustrated," it carries all there is to see in what it writes about. Through the laboratory, the text and the spectacle of the world end up having the same character.

9. But the last advantage is the greatest. The two-dimensional character of inscriptions allow them to *merge with geometry.* As we saw for perspective, space on paper can be made continuous with three-dimensional space. The result is that we can work on paper with rulers and numbers, but still manipulate three-dimensional objects "out there" (Ivins, 1973). Better still, because of this optical consistency, everything, no matter where it comes from, can be converted into diagrams and numbers, and combinations of numbers and tables can be used which are still easier to handle than words or silhouettes (Dagognet, 1973). You cannot measure the sun, but you can measure a photograph of the sun with a ruler. Then the number of centimeters read can easily migrate through different scales, and provide solar masses for completely different objects. This is what I call, for want of a better term, the *second-degree*

advantage of inscriptions, or the surplus-value that is gained through their capitalization.

These nine advantages should not be isolated from one another and should always be seen in conjunction with the mobilization process they accelerate and summarize. In other words, every possible innovation that offers any of these advantages will be selected by eager scientists and engineers: new photographs, new dyes to color more cell cultures, new reactive paper, a more sensitive physiograph, a new indexing system for librarians, a new notation for algebraic function, a new heating system to keep specimens longer. History of science is the history of these innovations. The role of the mind has been vastly exaggerated, as has been that of perception (Arnheim, 1969). An average mind or an average man, with the same perceptual abilities, within normal social conditions, will generate totally different output depending on whether his or her average skills apply to the confusing world or to inscriptions.

It is especially interesting to focus on the ninth advantage, because it gives us a way to make "formalism" a more mundane and a more material reality. To go from "empirical" to "theoretical" sciences is to go from slower to faster mobiles, from more mutable to less mutable inscriptions. The trends we studied above do not break down when we look at formalism but, on the contrary, increase fantastically. Indeed, what we call formalism is *the acceleration of displacement without transformation*. To grasp this point, let us go back to section 2. The mobilization of many resources through space and time is essential for domination on a grand scale. I proposed to call these objects that allow this mobilization to take place "immutable mobiles." I also argued that the best of these mobiles had to do with written, numbered, or optically consistent paper surfaces. But I also indicated, though without offering an explanation, that we had to deal with *cascades* of ever more simplified and costlier inscriptions. This ability to form a cascade has now to be explained because gathering written and imaged resources in one place, even with two-way connections, does *not* by itself guarantee any superiority for the one who gathers them. Why? Because the gatherer of such traces is immediately swamped in them. I showed such a phenomenon at work in Guillemin's laboratory; after only a few days of letting the instruments run, the piles of printout were enough to boggle the mind (Latour and Woolgar, 1979, chap. 2). The same thing happened to Darwin after a few years of collecting specimens with the *Beagle;* there were so many crates that Darwin was almost squeezed out of his house. So by themselves the inscriptions do *not* help a location to become a center that dominates the rest of the world. Something has to be done to the inscriptions which is similar to what the inscriptions do to the "things," so that at the end a few elements can manipulate all the others on a vast scale. The same deflating strategy we used to show how "things" were turned into paper, can show how paper is turned into *less* paper.

Let us take as example "the effectiveness of Galileo's work" as it is seen by Drake (1970). Drake does indeed use the word formalism to designate what Galileo is able to do that his predecessors were not. But what is described is more interesting than that. Drake compares the diagrams and commentaries of Galileo with those two older scholars, Jordan and Stevin. Interestingly, in Jordan's demonstration "the physical element is, as you see, brought in as an afterthought to the geometry, by main force as it were" (1970, 103). With Simon Stevin's diagram, this is the opposite: "The previous situation is reversed; geometry is eliminated in favor of pure mechanical intuition" (1970, 103). So, what seems to happen is that Galileo's two predecessors could not visually accommodate the problem on a paper surface and see the result simultaneously as both geometry and physics. A simple change in the geometry used by Galileo allows him to connect many different problems, whereas his two predecessors worked on disconnected shapes over which they had no control:

> Galileo's way of merging geometry and physics became apparent in his proof of the same theorem in his early treatise on motion dating from 1590. The method itself suggested to him not only many corollaries but successive improvements of the proof itself and further physical implications of it. (Drake, 1970, 104)

This ability to connect might be located in Galileo's mind. In fact what gets connected are three different visual horizons held synoptically because the surface of paper is considered as geometrical space:

> you see how the entire demonstration constitutes a *reduction* of the problem of equilibrium on inclined planes to the lever, which in itself removes the theorem from the isolation in which it stood before. (Drake, 1970, 106)

This innocuous term "removing from isolation" is constantly used by those who talk of theories. No wonder. If you just hold Galileo's diagram, you hold three domains; when you hold the others, only one. The holding allowed by a "theory" is no more mysterious (and no less) than the holding of armies, or of stocks, or of positions in space. It is fascinating to see that Drake explains the efficiency of Galileo's connection in terms of his creation of a geometrical medium is which geometry and physics merge. This is a much more material explanation than Koyré's idealist one, although the "matter" in Drake's rendering is a certain type of inscription on papers and certain ways of looking at it.

Similar tactics that use diagrams in order to establish rapid links between many unrelated problems are documented by cognitive psychologists. Herbert Simon (1982) compares the tactics of experts and novices in drawing diagrams when they are questioned about simple physical problems (pumps, water flows, and so on). The crucial difference between experts and novices is exactly the same as that pointed out by Drake:

> the crucial thing that appeared in the expert behaviour was that the formulation from the initial and the final condition was assembled in such a way that the relations between them and hence the answer could essentially be read off from it [the diagram]. (Simon, 1982, 169).

With this question in mind, one is struck by the metaphors "theoreticians" use to celebrate and rank theories.[12] The two main sets of metaphors insist respectively upon increased mobility and increased immutability. Good theories are opposed to bad ones or to "mere collections of empirical facts" because they provide "easy access to them." Hankel, for instance, criticizes Diophanus in the words that a French civil engineer would use to denigrate the Nigerian highway system:

> Any question requires a quite special method, which after will not serve even for the most closely allied problems. It is on that accord difficult for a modern mathematician even after studying one hundred Diophantine solutions, to solve the 101st problem; and if we have made the attempt, and after some vain endeavours read Diophantus' own solution, we shall be astonished to see how suddenly he leaves the broad highroad, dashes into a side path and with a quick turn reaches the goal . . . (cited in Bloor, 1976:102)

The safe path of science, as Kant would say, is not the same for the Greeks, for the Bororos and for us; but neither are the systems of transportation identical. One could object that these are only metaphors. Yes, but the etymology of *metaphoros* is itself enlightening. Precisely, it means displacement, transportation, transfer. No matter if they are mere images, these metaphors aptly *carry* the obsession of theoreticians for easy transportation and rapid communication. A more powerful theory, we submit, is one that with fewer elements and fewer and simpler transformations makes it possible to get at every other theory (past and future). Every time a powerful theory is celebrated it is always possible to rephrase this admiration in terms of the most trivial struggle for power: holding this place allows me to hold all the others. This is the problem we have encountered right through this paper: how to assemble many allies in one place (Latour, 1988b). Inscriptions allow *conscription!*

A similar link between ability to abstract and the practical work of mobilizing resources without transforming them is seen in much of cognitive science. In Piaget's tests, for instance, much fuss is made of water poured from a tall thin beaker into a short flat one. If the children say the water volume has changed, they are nonconserving. But as any laboratory observer knows, most of the phenomena depend upon which measure to read, or which to believe in case of discrepancy. The shift from nonconserving to conserving might not be a modification in cognitive structure, but a shift in indicators: read the height of the water in the first beaker and believe it *more* than the reading from the flat beaker. The notion of "volume" is *held* between the calibrated beakers exactly like Guillemin's endorphin is held between several peaks from at least five different instruments. In other words, Piaget is asking his children to do a laboratory experiment comparable in difficulty to that of the average Nobel Prize winner. If any shift in thinking occurs, it has nothing to do with the mind, but with the manipulation of the laboratory setting. Out of this setting no answer can be offered on volume. The best proof of this is that without industrially calibrated beakers Piaget himself would be totally unable to decide what is conserved (see also Cole and Scribner, 1974:last chapter). So again, most of what we grant a priori to "higher cognitive functions" might be concrete tasks done with new calibrated, graduated, and written objects. More generally, Piaget is obsessed with conservation and displacement through space without alteration (Piaget and Garcia, 1983). Thinking is tantamount to acquiring the ability to move as fast as possible while conserving as much of the pattern as possible. What Piaget takes as the logic of the psyche, is the very logic of mobilization and immutability which is so peculiar to our scientific societies, when they want to produce hard facts to dominate on a large scale. No wonder that all these "abilities" to move fast in such a world get better with schooling![13]

We now come closer to an understanding of the matter that constitutes formalism. The point of departure is that we are constantly hesitating between several often contradictory indications from our senses. Most of what we call "abstraction" is in practice the belief that a written inscription must be believed more than any contrary indications from the senses.[14] Koyré, for instance, has shown that Galileo believed in the inertia principle on mathematical grounds even against the contrary evidences offered to him not only by the Scriptures, but also by the senses. Koyré claims that this rejection of the senses was due to Galileo's Platonist philosophy. This might be so. But what does it mean practically? It means that faced with many contrary indications, Galileo, in the last instance, believed *more* in the triangular diagram for calculating the law of falling bodies, than any *other* vision of falling bodies (Koyré, 1966:147). When in doubt, believe the inscriptions, written in mathematical terms, *no matter* to what absurdities this might lead you.[15]

After Eisenstein's magisterial reworking of the Book of Nature argument, and Alper's redefinition of "visual culture," the ethnography of abstraction might be easier: What is this society in which a written, printed, mathematical form has greater credence, in case of doubt, than anything else: common sense, the senses other than vision, political authority, tradition, and even the Scriptures? It is obvious that this feature of society is overdetermined since it can be found in the written Law (Clanchy, 1979); in the biblical exegesis of the Holy Scriptures and in the history of geometry (Husserl, 1954; Derrida, 1967; Serres, 1980). Without this peculiar tendency to privilege what is written, the power of inscription would be entirely lost, as Edgerton hints in his discussion of Chinese diagrams. No matter how beautiful, rich, precise, or realistic inscriptions may be, no one would believe what they showed, if they could be contradicted by other evidence of local, sensory origin or pronouncements of the local authorities. I feel that we would make a giant step forward if we could relate this peculiar feature of our culture with the requirement of mobilization I have outlined several times. Most of the "domain" of cognitive psychology and epistemology does not exist but is related to this strange anthropological puzzle: a training (often in schools) to manipulate written inscriptions, to array them in cascades, and to believe the last one on the series more than any evidence to the contrary. It is in the description of this training that the anthropology of geometry and mathematics should be decisive (Livingston, 1986; Lave, 1985, 1986; Serres, 1982).

5. Paperwork

There are two ways in which the visualization processes we are all interested in may be ignored; one is to grant to the scientific mind that which should be granted to the hands, to the eyes, and to the signs; the other is to focus exclusively on the signs *qua* signs, without considering the mobilization of which they are but the fine edge. All innovations in picture making, equations, communications, archives, documentation, instrumentation, argumentation, will be selected for or against depending on how they simultaneously affect either inscription or mobilization. This link is visible not only in the empirical sciences, not only in the (former) realm of formalism, but also in many "practical" endeavors from which science is often unduly severed.

In a beautiful book, Booker retraces the history of engineering drawings (1982). Linear perspective (see above) progressively "changed the concept of pictures from being just representation to that of their being projections onto planes" (p. 31). But perspective still depended on the observer's position, so the objects could not really be moved everywhere without corruption. Desargues's and Monge's works

> helped to change the "point of view" or way of looking at things mentally. In place of the imaginary lines of space—so difficult to conceive clearly—which were the basis of perspective at that time, projective geometry allowed perspective to be seen in terms of solid geometry. (Booker, 1982:34)

With descriptive geometry, the observer's position becomes irrelevant. "It can be viewed and photographed from any angle or projected onto any plane—that is, distorted—and the result remains true" (p. 35). Booker and still better Baynes and Push (1981) in a splendid book (see also Deforges, 1981) show how *a few* engineers could *master* enormous machines that did not yet exist. These feats cannot be imagined without industrial drawings. Booker, quoting an engineer, describes the change of scale that allows the few to dominate the many:

> A machine that has been drawn is like an ideal realisation of it, but in a material that costs little and is easier to handle than iron or steel. . . . If everything is first well thought out, and the essential dimensions determined by calculations or experience, the plan of a machine or installation of machines can be quickly put on paper and the whole thing as well as the detail can then most conveniently be submitted to the severest criticism. . . . If at first there is doubt as to which of various possible arrangements is the most desirable then they are all sketched, compared with one another and the most suitable can easily be chosen. (Booker, 1982, 187)

Industrial drawing does not only create a paper world that can be manipulated as if in three dimensions. It also creates a common place for many other inscriptions to come together; margins of tolerance can be inscribed on the drawing, the drawing can be used for economic calculation, for defining the tasks to be made, or for organizing the repairs and the sales.

> But drawings are of the utmost importance not only for planning but also for execution since by means of them the measurements

> and proportions of all the parts can be so sharply and definitely determined from the beginning that when it comes to manufacture it is only necessary to imitate in the materials used for construction exactly what is shown in the drawing.
>
> Every part of the machine can in general be manufactured independently of every other part; it is therefore possible to distribute the entire work among a great number of workers. . . . No substantial errors can arise in work organised in this manner and if it does happen that on a rare occasion a mistake has been made it is immediately known with whom the blame lies. (Booker, 1982, 188)

Realms of reality that seem far apart (mechanics, economics, marketing, scientific organization of work) are inches apart, once flattened out onto the same surface. The accumulation of drawings in an optically consistent space is, once again, the "universal exchanger" that allows work to be planned, dispatched, realized, and responsibility to be attributed.[16]

The connective quality of written traces is still more visible in the most despised of all ethnographic objects: the file or the record. The "rationalization" granted to bureaucracy since Hegel and Weber has been attributed by mistake to the "mind" of (Prussian) bureaucrats. It is all in the files themselves. A bureau is, in many ways, and more and more every year, a small laboratory in which many elements can be connected together just because their scale and nature has been averaged out: legal texts, specifications, standards, payrolls, maps, surveys (ever since the Norman conquest, as shown by Clanchy, 1979). Economics, politics, sociology, hard sciences, do not come into contact through the grandiose entrance of "interdisciplinarity" but through the back door of the *file*. The "cracy" of bureaucracy is mysterious and hard to study, but the "bureau" is something that can be empirically studied, and which explains, because of its structure, why some power is given to an average mind just by looking at files: domains which are far apart become literally inches apart; domains which are convoluted and hidden become flat; thousands of occurrences can be looked at synoptically. More importantly, once files start being gathered everywhere to ensure some two-way circulation of immutable mobiles, they can be arrayed in a cascade: files of files can be generated and this process can be continued until a few men consider millions as if they were in the palms of their hands. Common sense ironically makes fun of these "gratte-papiers" or "paper shufflers," and often wonders what all this "red tape" is for; but the same question should be asked of the rest of science and technology. In our cultures "paper shuffling" is the source of an essential power, that constantly escapes attention since its materiality is ignored.

McNeill, in his fundamental book *The Pursuit of Power* (1982), uses this ability to distinguish Chinese bureaucracy from that of the Occident. Accumulation of records and ideograms make the Chinese Empire possible. But there is a major drawback with ideograms; once gathered you cannot array them in a cascade in such a way that thousands of records can be turned in one, that is literally "punctualized" through geometrical or mathematical skills. So here again, if we keep both the quality of the signs and the mobilization process in focus, we may understand why careful limits have been put in the past to the growth of the Chinese imperium, and why these limits to the mobilization of resources on a grand scale have been broken in Europe. It is hard to overestimate the power that is gained by concentrating files written in a homogeneous and combinable form (Wheeler, 1969; Clanchy, 1979).

This role of the bureaucrat qua scientist qua writer and reader, is always misunderstood because we take for granted that there exist, somewhere in society, macro-actors that naturally dominate the scene: Corporation, State, Productive Forces, Cultures, Imperialism, "Mentalités," etc. Once accepted, these large entities are then used to explain (or to not explain) "cognitive" aspects of science and technology. The problem is that these entities could not exist at all without the construction of long networks in which numerous faithful records circulate in both directions, records which are, in turn, summarized and displayed to convince. A "state," a "corporation," a "culture," or an "economy" are the result of a punctualization process that obtains a few indicators out of many traces. In order to exist these entities have to be *summed up* somewhere (Chandler, 1977; Beniger, 1986). Far from being the key to the understanding of science and technology, these entities are the very things a new understanding of science and technology should explain. The large-scale actors to which sociologists of science are keen to attach "interests" are immaterial in practice as long as precise mechanisms to explain their origin or extraction and their changes of scale have not been proposed.

A man is never much more powerful than any other—even from a throne; but a man whose eye dominates records through which some sort of connections are established with millions of others may be said to *dominate*. This domination, however, is not a given but a slow construction and it can be corroded, interrupted, or destroyed if the records, files, and figures are immobilized, made more mutable, less readable, less combinable, or unclear when displayed. In other words, the *scale* of an actor is not an absolute term but a relative one that varies with the ability to produce, capture, sum up, and interpret information about other places and times (Callon and Latour, 1981). Even the very notion of scale is impossible to understand without an inscription or a map in mind. The "great man" is a little man looking at a good map. In Mercator's frontispiece Atlas is transformed from a god who carries the world into a scientist who holds it in his hand!

Since the beginning of this presentation on how to draw things together, I have been recasting the simple question of power: how the few may dominate the many. After McNeill's major reconceptualization of the history of power in terms of mobilization, this age-old question of political philosophy and sociology can be rephrased in another way: how can distant or foreign places and times be gathered in one place in a form that allows all the places and times to be presented at once, and which allows orders to move *back* to where they came from? Talking of power is an endless and mystical task; talking of distance, gathering, fidelity, summing up, transmission, etc., is an empirical one, as has been illustrated in a recent study by John Law of the Portuguese spice road to India (1986). Instead of using large-scale entities to explain science and technology as most sociologists of science do, we should start from the inscriptions and their mobilization and see how they help small entities to become large ones. In this shift from one research program to another, "science and technology" will cease to be the mysterious cognitive object to be explained by the social world. It will become one of the main sources of power (McNeill, 1982). To take the existence of macro-actors for granted without studying the material that makes them "macro," is to make both science and society mysterious. To take the fabrication of various scales as our main center of interest is to place the practical means of achieving power on a firm foundation (Cicourel, 1981). The Pentagon does not *see* more of the Russians' strategy than Guillemin does his endorphin. They simply put faith in superimposed traces of various quality, opposing some to others, retracing the steps of those that are dubious, and spending billions to create new branches of science and technology that can accelerate the mobility of traces, perfect their immutability, enhance readability, ensure their compatibility, quicken their display: satellites, networks of espionage, computers, libraries, radioimmunoassays, archives, surveys. They will never see more of the phenomena than what they can build through these many immutable mobiles. This is *obvious,* but rarely *seen.*

If this little shift from a social/cognitive divide to the study of inscriptions is accepted, then the importance of *metrology* appears in proper light. Metrology is the scientific organization of stable measurements and standards. Without it no measurement is stable enough to allow either the homogeneity of the inscriptions or their return. It is not surprising then to learn that metrology costs up to three times the budget of all research and development, and that this figure is for only the first elements of the metrological chain (Hunter, 1980). Thanks to metrological organization the basic physical constants (time, space, weight, wavelength) and many biological and chemical standards may be extended "everywhere" (Zerubavel, 1982; Landes, 1983). The universality of science and technology is a cliché of epistemology but metrology is the practical achievement of this mystical universality. In practice it is costly and full of holes (see Cochrane, 1966 for the history of the Bureau of Standards). Metrology is only the official and primary component of an ever-increasing number of measuring activities we all have to undertake in daily life. Every time we look at our wristwatch or weigh a sausage at the butcher's shop, every time applied laboratories measure lead pollution, water purity, or control the quality of industrial goods, we allow more immutable mobiles to reach new places. "Rationalization" has very little to do with the reason of bureau- and technocrats, but has a lot to do with the maintenance of metrological chains (Uselding, 1981). This building of long networks provides the stability of the main physical constants, but there are many other metrological activities for less "universal" measures (polls, questionnaires, forms to fill in, accounts, tallies).

There is one more domain into which this ethnography of inscription could bring some "light." I want to talk about it since at the beginning of this review I rejected dichotomies between "mentalist" and "materialist" explanations. Among the interesting immutable mobiles there is one that has received both too little and too much attention: money. The anthropology of money is as complicated and entangled as that of writing, but one thing is clear. As soon as money starts to circulate through different cultures, it develops a few clearcut characteristics: it is mobile (once in small pieces), it is immutable (once in metal), it is countable (once it is coined), combinable, and can circulate from the things valued to the center that evaluates and back. Money has received too much attention because it has been thought of as something special, deeply inserted in the infrastructure of economies, whereas it is just one of the many immutable mobiles necessary if one place is to exercise power over many other places far apart in space and time. As a type of immutable mobile *among others* it has, however, received too little attention. Money is used to code all states of affairs in exactly the way that La Pérouse coded all places by longitude and latitude (actually, in his log book La Pérouse registered both the places on the map and the values of each good as if it were to be sold in some other place). In this way, it is possible to accumulate, to count, to display, and to recombine all the states of affairs. Money is neither more nor less "material" than mapmaking, engineering drawings, or statistics.

Once its ordinary character is recognized, the "abstraction" of money can no longer be the object of a fetish cult. For instance, the importance of the art of accounting both in economies and science falls nicely into place. Money is not interesting as such but as one type of immutable mobile that links goods and places; so it is no wonder if it quickly merges with other written inscriptions such as figures, columns, and double-entry bookkeeping (Roover, 1963). No wonder if, through accounting, it is possible to gain more just by recombining numbers (Braudel, 1979, especially vol. 3; Chandler, 1977). Here again, too much emphasis should not be placed on the visualization of numbers per se; what should really be stressed is the cascade of mobile inscriptions that end up in an account, which is, literally, the only thing that *counts*. Exactly as with any scientific inscription, in case of doubt the new accountant prefers to believe inscription, no matter how strange the consequences and counterintuitive the phenomena. The history of money is thus seized by the same trend as all the other immutable mobiles; any innovations that can accelerate money to enlarge its power of mobilization are kept: checks, endorsement, paper money, electronic money. This trend is not due to the development of capitalism. "Capitalism" is, on the contrary, an empty word as long as precise material instruments are not proposed to explain any capitalization at all, be it of specimens, books, information or money.

Thus, capitalism is not to be used to explain the evolution of science and technology. It seems to me that it should be quite the contrary. Once science and technology are rephrased in terms of immutable mobiles it might be possible to explain economic capitalism as another process of mobilization and conscription. What indicates this are the many weaknesses of money; money is a nice immutable mobile that circulates from one point to another but it carries very little with it. If the name of the game is to accumulate enough allies in one place to modify the belief and behavior of all the others, money is a poor resource as long as it is isolated. It becomes useful when it is combined with all the other inscription devices; then, the different points of the world become really transported in a manageable form to a single place which then becomes a *center*. Just as with Eisenstein's printing press, which is one factor that allows all the others to merge with one another, what counts is not the capitalization of money, but the capitalization of all compatible inscriptions. Instead of talking of merchants, princes, scientists, astronomers, and engineers as having some sort of relation with one another, it seems to me it would be more productive to talk about *"centers of calculation."* The currency in which they calculate is less important than the fact that they calculate only with inscriptions and mix together in these calculations inscriptions coming from the most diverse disciplines. The calculations themselves are less important than the way they are arrayed in cascades, and the bizarre situation in which the last inscription is believed more than anything else. Money per se is certainly not the universal standard looked for by Marx and other economists. This qualification should be granted to centers of calculation and to the peculiarity of written traces which makes rapid translation between one medium and another possible.

Many efforts have been made to link the history of science with the history of capitalism, and many efforts have been made to describe the scientist as a capitalist. All these efforts (including mine—Latour and Woolgar, 1979, chap. 5; Latour, 1984a) were doomed from the start, since they took for granted a division between mental and material factors, an artifact of our ignorance of inscriptions.[17] There is not a history of engineers, then a history of capitalists, then one of scientists, then one of mathematicians, then one of economists. Rather, there is a single history of these centers of calculation. It is not only because they look exclusively at maps, account books, drawings, legal texts, and files, that cartographers, merchants, engineers, jurists, and civil servants get the edge on all the others. It is because all these inscriptions can be superimposed, reshuffled, recombined, and summarized, and that totally new phenomena emerge, hidden from the other people from whom all these inscriptions have been exacted.

More precisely we should be able to explain, with the concept and empirical knowledge of these centers of calculation, how insignificant people working only with papers and signs become the most powerful of all. Papers and signs are incredibly weak and fragile. This is why explaining anything with them seemed so ludicrous at first. La Pérouse's map is not the Pacific, anymore than Watt's drawings and patents are the engines, or the bankers' exchange rates are the economies, or the theorems of topology are "the real world." This is precisely the paradox. By working on papers alone, on fragile inscriptions that are immensely less than the things from which they are extracted, it is still possible to dominate all things and all people. What is insignificant for all other cultures becomes the most significant, the only significant aspect of reality. The weakest, by manipulating inscriptions of all sorts obsessively and exclusively, become the strongest. This is the view of power we get at by following this theme of visualization and cognition in all its consequences. If you want to understand what draws *things* together, then look at what *draws* things *together.*

Notes

1. For instance, Levi-Strauss' divide between bricoleur and engineer or between hot and cold societies (1962); or Garfinkel's distinctions between

everyday and scientific modes of thought (1967); or Bachelard's many "coupures épistémologiques" that divide science from common sense, from intuition, or from its own past (1934, 1967); or even Horton's careful distinction between monster acceptance and monster avoidance (1977) or primary theories and secondary theories (1982).

2. Goody (1977) points to the importance of practical tasks in handling graphics (lists, dictionaries, inventories), and concludes his fascinating book by saying that "if we wish to speak of a 'savage mind' these are some of the instruments of its domestication" (p. 182). Cole and Scribner (1974) shift the focus from intellectual tasks to schooling practice; the ability to draw syllogisms is taken out of the mind and put into the manipulation of diagrams on paper. Hutchins (1980) does the opposite in transforming the "illogical" reasoning of the Trobriand islanders into a quite straightforward logic simply by adding to it the land use systems that give meaning to hitherto abrupt shifts in continuity. Eisenstein switches the enquiry from mental states and the philosophical tradition to the power of print (1979). Perret-Clermont (1979), at first one of Piaget's students, focuses her attention on the social context of the many test situations. She shows how "non-conserving" kids become conserving in a matter of minutes simply because other variables (social or pictoral) are taken into account. Lave has explored in pioneering studies how mathematical skills may be totally modified depending on whether or not you let people use paper and pencil (Lave, 1986, 1988; Lave, Murtaugh and De La Rocha, 1983). Ferguson has tried to relate engineering imagination to the abilities to draw pictures according to perspective rules and codes of shades and colors (1977): "It has been nonverbal thinking by and large that has fixed the outlines and filled in the details of our material surroundings. . . . Pyramids, cathedrals, and rockets exist not because of geometry, theory of structures or thermodynamics, but because they were first a picture—literally a vision—in the minds of those who built them" (p. 835) (See also Ferguson, 1985). These are some of the studies that put the deflating strategy I try to review here into practice.
3. A fact is harder or softer as a function of what happens to it in other hands later on. Each of us acts as a multi-conductor for the many claims that we come across: we may be uninterested, or ignore them, or be interested but modify them and turn them into something entirely different. Sometimes indeed we act as conductor and pass the claim along without further modification. (For this see Latour and Woolgar, 1979; Latour, 1984b.)
4. "Science and technology have advanced in more than direct ratio to the ability of men to contrive methods by which the phenomena which otherwise could be known only through the senses of touch, hearing, taste and smell, have been brought within the range of visual recognition and measurements and then become subject to that logical symbolization without which rational thought and analysis are impossible" (Ivins, 1973, 13).
5. "The most marked characteristics of European pictorial representation since the fourteenth century, have been on the one hand its steadily increasing naturalism and on the other its purely schematic and logical ex-

tension. It is submitted that both are due in largest part to the development and pervasion of methods which have provided symbols, repeatable in invariant forms, for representation of visual awareness and a grammar of perspective which made it possible to establish logical relations not only within the system of symbols but between that system and the forms and locations of the objects that it symbolizes" (Ivins, 1973, 12).

6. "Northern artists characteristically sought to represent by transforming the extent of vision onto their small, flat working surface. . . . It is the capacity of the picture surface to contain such a semblance of the world—an aggregate of views—that characterizes many pictures in the North" (Alpers, 1983, 51).
7. The proof that the *movement* comes first, for Eisenstein, lies in the fact that it entails exactly the opposite effects on the Scriptures. The accuracy of the medium reveals more and more inaccuracies in the message, which is soon jeopardized. The beauty of Eisenstein's construction resides in the way it obtains two opposite consequences from the same cause: science and technology accelerates; the Gospel becomes doubtful (Latour, 1983).
8. For instance, Mukerji portrays a geographer who hates the new geography books but has to cry his hate in print: "Ironically, Davis took his trip because he did not trust printed information to be as complete as oral accounts of experiences; but he decided to make the voyage after reading Dutch books on geography and produced from his travel another geographical/navigational text" (Mukerji, 1983, 114).
9. This is why I do not include in the discussion the large literature on the neurology of vision or on the psychology of perception (see for instance Block, 1981; de Mey, 1982). These disciplines, however important, make so much use of the very process I wish to study that they are as blind as the others to an ethnography of the crafts and tricks of the visualization.
10. "Un 'pouvoir d'écriture' se constitute comme une pièce essentielle dans les rouages de la discipline. Sur bien des points, il se modèle sur les méthodes traditionnelles de la documentation administrative mais avec des techniques particulières et des innovations importantes" (Foucault, 1975, 191).
11. These simple shifts are often transformed by philosophers into complete ruptures from common sense, into "coupures épistémologiques" as in Bachelard. It is not because of the empiricists' naïveté that one has to fall back on the power of theories to make sense of data. The focus on inscriptions and manipulation of traces is exactly midway between empiricism and Bachelard's argument on the power of theories.
12. A nice example is that of Carnot's thermodynamics studied by Redondi (1980). Carnot's know-how is not about building a machine but rather a diagram. This diagram is drawn in such a way that it allows one to move from one engine to any other, and indeed to nonexistent engines simply *drawn* on paper. Real three-dimensional steam engines are interesting but localized and cumbersome. Thermodynamics is to them what La Pérouse's map is to the islands of the Pacific. When going from one engine to the

theory or from one island to the map, you do not go from concrete to abstract, from empirical to theoretical, you go from one place that dominates no one, to another place that dominates all the others. If you grasp thermodynamics you grasp all engines (past, present and future—see Diesel). The question about theories is: who controls whom and on what scale.

13. A nice *a contrario* proof is provided by Edgerton's study of Chinese technical drawings (1980). He claims that Chinese artists have no interest in the figures or, more exactly, that they take figures not inside the perspective space on which an engineer can work and make calculations and previsions, but as *illustrations*. In consequence, all the links between parts of the machines become decorations (a complex part of the pump becomes, for instance, waves on a pond after a few copies!). No one would say that Chinese are unable to abstract, but it would not be absurd to say that they do not put their full confidence into writing and imaging.
14. In a beautiful article Carlo Ginzburg speaks of a "paradigm of the trace" to designate this peculiar obsession of our culture that he traces—precisely!—from Greek medicine, to Conan Doyle's detective story, through Freud's interest in lapsus and the detection of art forgeries (1980). Falling back, however, on a classical prejudice, Ginzburg puts physics and hard sciences aside from such a paradigm because, he contends, they do not rely on traces but on abstract, universal phenomena!
15. Ivins explains, for instance, that most Greek parallels in geometry do not meet because they are touched with the hands, whereas Renaissance parallels do meet since they are only seen on paper (1973:7). Jean Lave, in her studies of Californian grocery shoppers, shows that people confronted with a difficulty in their computation rarely stick to the paper and never put their confidence in what is written (Lave et al., 1983). *To do so* no matter how absurd the consequences requires still another set of peculiar circumstances related to laboratory settings, even if these are as Livingston says (1986) "flat laboratories." In one of his twelve or so origins of geometry Serres argues that having invented the alphabet and thus broken any connection between written shapes and the signified, the Greeks had to cope with pictorial representation. He argues that what we came to call formalism is an alphabetic text trying to describe visual diagrams: "Qu'est-ce que cette géométrie dans la pratique? Non point dans les 'idées' qu'elle suppose mais dans l'activité qui la pose. Elle est d'abord un art du dessin. Elle est ensuite un langage qui parle du dessin tracé que celui-ci soit présent ou absent" (Serres, 1980, 176).
16. The link between technical thinking and technical drawing is so close that scholars establish it even unwillingly. For instance, Bertrand Gille, when accounting for the creation of a new "système technique" in Alexandria during the Hellenistic period, is obliged to say that it is the availability of a good library and the gathering of a collection of scale models of all the machines previously invented, that transformed "mere practice" into techno-*logy* (1980). What makes the "système technique" a *system* is the synoptic vision of all the former technical achievements which are all taken

out of their isolation. This link is most clearly visible when an inscription device is hooked up to a working machine to make it comprehensible (Hills and Pacey, 1981; Constant, 1983). A nice rendering of the paper world necessary to make a computer real is to be found in Kidder (1981). "The *soul* of the machine" is a pile of paper. . . .

17. The *direction* we go to by asking such questions is quite different from those of either the sociology of science or the cognitive sciences (especially when they both try to merge as in de Mey's synthesis (1982)). Two recent attempts have been made to relate the fine structure of cognitive abilities to social structure. The first one uses Hesse's networks and Kuhn's paradigms (Barnes, 1982), the second Wittgenstein's "language games" (Bloor, 1983). These attempts are interesting but they still try to answer a question which the present review wishes to reject: how cognitive abilities are *related* to our societies. The question (and thus the various answers) accept the idea that the stuff society is made of is somehow different from that of our sciences, our images, and our information. The phenomenon I wish to focus on is slightly different from those revealed by Barnes and Bloor. We are dealing with a single ethnographic puzzle: some societies—very few indeed—are made by capitalizing on a larger scale. The obsession with rapid displacement and stable invariance, for powerful and safe linkages, is not a part of our culture, or "influenced" by social interests: *it is our culture*. Too often sociologists look for *indirect* relations between "interests" and "technical" details. The reason of their blindness is simple: they limit the meaning of "social" to society without realizing that the mobilizing of allies and, in general, the transformation of weak into strong associations, is what "social" also means. Why look for far-fetched relations when technical details of science talk directly of invariance, association, displacement, immutability, and so on? (Law, 1986; Latour, 1984b; Callon, Law, and Rip, 1986).

References

Alpers, S. (1983). *The Art of Describing: Dutch Art in the 17th Century*. Chicago: University of Chicago Press.
Arnheim, R. (1969). *Visual Thinking*. Berkeley: University of California Press.
Augé, M. (1975). *Theorie des Pouvoirs et Idéologie*. Paris: Hermann.
Bachelard, G. (1934). *Le Nouvel Esprit Scientifique*. Paris: PUF.
Bachelard, G. (1967). *La Formation de l'Esprit Scientifique*. Paris: Vrin.
Barnes, B. (1982). *T. S. Kuhn and Social Science*. London: Macmillan.
Baynes, K. and F. Push. (1981). *The Art of the Engineer*. Guildford, Sussex: Lutherword Press.
Beniger, J. R. (1986). *The Control Revolution*. Cambridge, MA: Harvard University Press.
Bertin, P. (1973). *Sémiologie Graphique*. Paris: Mouton.

Block, N., Ed. (1981). *Imagery*. Cambridge, MA: The MIT Press.

Bloor, D. (1976). *Knowledge and Social Imagery*. London: Routledge.

Bloor, D. (1983). *Wittgenstein and the Social Theory of Knowledge*. London: Macmillan.

Booker, P. J. (1982). *A History of Engineering Drawing*. London: Northgate Publishing Co.

Bourdieu, P. (1972). *Esquisse d'une Théorie de la Pratique*. Genève: Droz.

Braudel, L. (1979). *Civilisation Matérielle et Capitalisme*. Paris: Armand Colin.

Callon, M. and B. Latour, (1981). "Unscrewing the big Leviathan." In K. Knorr and A. Cicourel (Eds.), *Toward an Integration of Micro and Macro Sociologies*. London: Routledge.

Callon, M., J. Law, and A. Rip, Eds. (1986). *Qualitative Scientometrics: Studies in the Dynamic of Science*. London: Macmillan.

Chandler, A. (1977). *The Visible Hand*. Cambridge, MA: Harvard University Press.

Cicourel, A. (1981). "Notes on the integration of micro and macro levels." In K. Knorr and A. Cicourel (Eds.), *Toward an Integration of Micro and Macro Sociologies*. London: Routledge.

Clanchy, M. T. (1979). *From Memory to Written Records 1066–1300*. Cambridge, MA: Harvard University Press.

Cochrane, R. X. (1966). *Measure for Progress: A History of the National Bureau of Standards*. Washington, D.C.: U.S. Bureau of Commerce.

Cole, J., and S. Scribner (1974). *Culture and Thought: A Psychological Introduction*. New York: John Wiley and Sons.

Constant, E. W. (1983). "Scientific theory and technological testability: science, dynamometer and water turbine in the 19th century." *Technology and Culture* 24(2):183–198.

Dagognet, F. (1969). *Tableaux et Langages de la Chimie*. Paris: Le Seuil.

Dagognet, F. (1973). Ecriture et Iconographie. Paris: Vrin.

Deforges, Y. (1981). *Le Graphisme Technique*. Le Creusot: Editions Champs-Vallon.

de Mey, M. (1982). *The Cognitive Paradigm*. Dordrecht: Reidel.

Derrida, J. (1967). *De la Gramatologie*. Paris: Minuit.

Drake, S. (1970). *Galileo Studies*. Ann Arbor: University of Michigan Press.

Edgerton, S. (1976). *The Renaissance Discovery of Linear Perspective*. New York: Harper and Row.

Edgerton, S. (1980). "The Renaissance artist as a quantifier." In M. A. Hagen (Ed.), *The Perception of Pictures*, vol. 1. New York: Academic Press.

Eisenstein, E. (1979). *The Printing Press as an Agent of Change*. Cambridge: Cambridge University Press.

Fabian, J. (1983). *Time and the Other: How Anthropology Makes Its Object*. New York: Columbia University Press.

Ferguson, E. (1977). "The mind's eye: nonverbal thought in technology." *Science* 197:827ff.

Ferguson, E. (1985). "La Fondation des machines modernes: des dessins." In B. Latour (Ed.), *Les 'Vues' de l'Esprit*, special issue of *Culture Technique*.
Foucault, M. (1963). *Naissance de la Clinique: Une Archéologie du Regard Médical*. Paris: PUF.
Foucault, M. (1966). *Les Mots et Les Choses*. Paris: Gallimard.
Foucault, M. (1975). *Surveiller et Punir*. Paris: Gallimard.
Fourquet, M. (1980). *Le Comptes de la Puissance*. Paris: Encres.
Garfinkel, H. (1967). *Studies in Ethnomethodology*. Englewood Cliffs, NJ: Prentice-Hall.
Gille, B. (1980). *Les Ingénieurs Grecs*. Paris: Le Seuil.
Ginzburg, C. (1980). "Signes, traces, pistes." *Le Débat* 6:2–44.
Goody, J. (1977). *The Domestication of the Savage Mind*. Cambridge: Cambridge University Press.
Hagen, M. A. (1980). *The Perception of Pictures*, Tome I and II. New York: Academic Press.
Hanson, N. R. (1962). *Perception and Discovery: An Introduction to Scientific Inquiry*. San Francisco: W. H. Freeman.
Havelock, E. B. (1980). *Aux Origines de la Civilisation Écrite en Occident*. Paris: Maspero.
Hills, R. and A. J. Pacey (1982). "The measurement of power in early steam dream textile mills." *Technology and Culture* 13(1):25ff.
Hollis, M. and S. Lukes, Eds. (1982). *Rationality and Relativism*. Oxford: Blackwell.
Horton, R. (1977). "African thought and western science." In B. Wilson (Ed.), *Rationality*, Oxford: Blackwell.
Horton, R. (1982). "Tradition and modernity revisited." In M. Hollis and S. Lukes eds., *Rationality and Relativism*. Oxford: Blackwell.
Hughes, T. (1979). "The system-builders." *Technology and Culture* 20(1): 124–161.
Hunter, P. (1980). "The national system of scientific measurement." *Science* 210:869–874.
Husserl, E. (1954/1962). L'*Origine de la Géométrie*. Paris: PUF.
Hutchins, E. (1980). *Culture and Inference: A Trobriand Case Study*. Cambridge, MA: Harvard University Press.
Ivins, W. M. (1953). *Prints and Visual Communications*. Cambridge, MA: Harvard University Press.
Ivins, W. M. (1973). *On the Rationalization of Sight*. New York: Plenem Press.
Kidder, T. (1981). *The Soul of a New Machine*. London: Allen Lane.
Koyré, A. (1966). *Etudes Galiléennes*. Paris: Hermann.
Knorr, K. (1981). *The Manufacture of Knowledge*. Oxford: Pergamon Press.
Knorr, K. and A. Cicourel, Eds. (1981). *Toward an Integration of Micro and Macro Sociologies*. London: Routledge.
Landes, D. (1983). *Revolution in Time: Clocks and the Making of the Modern World*. Cambridge, MA: Harvard University Press.
La Pérouse, J. F. de (n.d.). *Voyages Autour du Monde*. Paris: Michel de l'Ormeraie.

Latour, B. (1983). "Comment redistribuer le grand partage?" *Revue Internationale de Synthèse* 104(110):202–236.
Latour, B. (1984a). "Le dernier des capitalistes sauvages, interview d'un biochimiste." *Fundamenta Scientiae* 4(3/4):301–327.
Latour, B. (1984b). *Le Microbes: Guerre et Paix suivi des Irréductions*. Paris: A.M. Métaiiié.
Latour, B. (1985). *Les 'Vues' de l'Esprit*, special issue of *Culture Technique*.
Latour, B. (1987). *Science in Action*. Cambridge, MA: Harvard University Press.
Latour, B. (1988a). *The Pasteurization* of France. Cambridge, MA: Harvard University Press.
Latour, B. (1988b). *A Relativistic Account of Einstein's Relativity*. Cambridge, MA: Harvard University Press.
Latour, B., and F. Bastide (1985). "Science-fabrication." In M. Callon, J. Law, and A. Rip (Eds.), *Qualitative Scientometrics: Studies in the Dynamic of Science*. London: Macmillan.
Latour, B., and S. Woolgar (1979/1986). *Laboratory Life: The Social Construction of Scientific Facts*. London: Sage. (1986, 2nd edition; Princeton: Princeton University Press.)
Lave, J. (1986). "The values of quantification." In J. Law Ed., *Power, Action and Belief*, London: Routledge.
Lave, J. (1988). *Cognition in Practice*. Cambridge, UK: Cambridge University Press.
Lave, J., M. Murtaugh, and O. De La Rocha (1983). "The dialectic constitution of arithmetic practice." In B. Rogoff and J. Lave (Eds.), *Everyday Cognition: Its Development in Social Context*. Cambridge, MA: Harvard University Press.
Law, J. (1983). "Enrôlement et contre-enrôlement: les luttes pour la publication d'un article scientifique." *Social Science Information* 22(2):237–251.
Law, J. (1985). "Les textes et leurs allies." In B. Latour (Ed.), *Les 'Vues' de l'Esprit*, special issue of *Culture Technique*.
Law, J. (1986). "On the methods of long-distance control: vessels, navigations and the Portuguese route to India." In J. Law (Ed.), *Power, Action and Belief*. London: Routledge.
Leroi-Gourhan, A. (1964). *Le Geste et la Parole*. Paris: Albin Michel.
Levi-Strauss, C. (1962). *La Pensée Sauvage*. Paris: Plon.
Livingston, E. (1986). *An Ethnomethodological Investigation of the Foundations of Mathematics*. London: Routledge & Kegan Paul.
Lynch, M. (1985a). "Discipline and the material form of images: an analysis of scientific visibility." *Social Studies of Science*. 15:37–66.
Lynch, M. (1985b). *Art and Artifact in Laboratory Science*. London: Routledge & Kegan Paul.
McNeill, W. (1982). *The Pursuit of Power, Technology, Armed Forces and Society Since A.D. 1000*. Chicago: University of Chicago Press.
Mukerji, S. (1983). *From Graven Images: Patterns of Modern Materialism*. New York: Columbia University Press.

Mukerji, S. (1985). "Voir le pouvoir." In B. Latour (Ed.), *Les 'Vues' de l'Esprit,* special issue of *Culture Technique.*

Ong, W. (1971). *Rhetoric, Romance and the New Technology.* Ithaca, NY: Cornell University Press.

Ong, W. (1982). *Orality and Literacy: The Technologizing of the Word.* London: Methuen.

Perret-Clermont, A. N. (1979). *La Construction de l'Intelligence dans l'Intéraction Sociale.* Berne: Peter Lang.

Piaget, J., and R. Garcia (1983). *Psychogenèse et Histoire des Sciences.* Paris: Flammarion.

Pinch, T. (1985). "Toward an analysis of scientific observations: the externality of evidential significance of observational reports in physics." *Social Studies of Science* 15:3–37.

Redondi, P. (1980). *L'accueil des Idées de Sadi Carnot: de la Légende à l'Histoire.* Paris: Vrin.

Roover, R. de (1963). *The Rise and Decline of the Medici Bank.* Cambridge, MA: Harvard University Press.

Rudwick, M. (1976). "The emergence of a visual language for geological science: 1760–1840." *History of Science* 14:148–195.

Serres, M. (1980). *Le Passage du Nord-Oeust.* Paris: Minuit.

Serres, M. (1982). *Hermes.* Baltimore, MD: Johns Hopkins University Press.

Shapin, S. (1984). "Pump and circumstance: Robert Boyle's literary technology." *Social Studies of Science* 14:481–521.

Simon, H. (1982). "Cognitive processes of experts and novices." *Cahiers de la Fondation Archives Jean Piaget* 2(3):154–178.

Sohn-Rethel, A. (1978). *Manual and Intellectual Labor: A Critique of Epistemology.* London: Macmillan.

Star, S. L. (1983). "Simplification in scientific work: an example from neuroscience research." *Social Studies of Science* 13:205–228.

Star, S. L. and J. R. Griesemer (1989). "Institutional ecology, 'translations,' and boundary objects: Amateurs and professionals in Berkeley's Museum of Vertebrate Zoology, 1907–1939." *Social Studies of Science* 19:387–420.

Uselding, P. (1981). "Measuring techniques and manufacturing practice." In O. Mayr and R. Post (Eds.), *Yankee Enterprise.* Washington, D.C.: Smithsonian Institute Press.

Wheeler, J. (1969). *On Records Files and Dossiers in American Life.* New York: Russell Sage Foundation.

Zerubavel, E. (1982). "The standardization of time: a sociohistorical perspective." *American Sociological Review* 88(1):1–29.

Representation and the realist-constructivist controversy

PAUL TIBBETTS
Department of Philosophy, University of Dayton, Dayton, OH 45469, USA

1. Introductory remarks: The realist constructivist debate

There are at least three issues surrounding the problem of representation: (1) The representational device or RD (e.g., maps, electrical diagrams, chemical formulae, models, etc.) and the extent to which such RDs are socially constructed, interpreted and deployed; (2) the ontological status of the represented object (RO); and (3) questions concerning the accuracy with which (1) represents (2). For the realist, RDs ultimately denote some independently existing non-cognitive structure or process. Realists recognize the constructivist dimension to RDs and the extent to which such devices and their use are defined by inquirer-contingent criteria, though they insist that RDs ultimately have to map onto some inquirer-independent, real-world properties. For realists, if this were not the case then RDs would represent nothing at all and therefore the data points provided by RDs would be unintelligible. (For variations on the realist position see: Bhaskar, 1978; Jarvie, 1983, 1984; Laudan, 1977, 1981; and Popper, 1972.)

One traditional problem with such realist accounts concerns just what it is RDs represent (issue (2) above). A pragmatic response to this query is that this is an issue best left to metaphysicians and those concerned with questions of ontology. What *does* matter – on this line of thinking – are the data points, particularly those points that: (i) are consistent with theory, (ii) are theoretically interesting, (iii) occasion theoretical revision or extension, and (iv) have heuristic value for further research. After all, why would the realist – or anyone else for that matter – be remotely interested in the relation between RDs and ROs, and the data points the latter occasion in the former, unless there were some theoretical or pragmatic significance associated with RDs? And, continuing this line of response, neither can relativists and constructivists in the sociology of science ignore the fact that RDs do generate reams of non-random data points.

Of course, constructivists counter that what constitutes theoretically significant as against non-significant, or non-random as against random, data points requires reference to epistemic criteria posited by a given community of inquirers. Such criteria include theoretical consistency, heuristic value, predictive accuracy, and Lakatos' (sophisticated) falsifiability. Apart from such negotiated and therefore contingent evaluative criteria, estimates of significance or non-significance are unresolvable in principle. (Cf. Barnes and Edge, 1982; Bohme, 1977; Collins, 1983a, 1983b, 1985; and Knorr-Cetina and Mulkay, 1983.) Still, realists can counter, such socially-constructed epistemic criteria must be criteria for evaluating *something*. And so the issue bounces back and forth between realist and constructivist. Understandably, a number of sociologists of science have increasingly turned their attention to this debate between realists' and constructivists' accounts of linguistic and non-linguistic representation in science. However, rather than pursuing this issue on a relatively abstract, philosophical level, these sociologists of science examine representational practices within the day-to-day context of inquiry and the interpretive controversy surrounding such practices. (Cf. Collins and Pinch, 1982; Garfinkel, Lynch and Livingston, 1981; Grenier, 1983; Knorr-Cetina, 1981; and Pickering, 1984.)

Nor is one necessarily committed to either a realist or a constructivist account in a mutually exclusive sense; elements of both in fact appear in most writings on the subject. The contrast between constructivism and realism is the *emphasis* respectively given – or not given – to the social contingencies surrounding RDs and associated evaluative criteria, and the supposed epistemic independence of the data points from such considerations.

In the remainder of this paper I examine some recent accounts in the sociology and philosophy of science literature regarding RDs, the RD–RO relation and the issue of data points, showing the extent to which realist and constructivist elements are co-present in these accounts. To focus our discussion, Section 2 below examines the account of RDs and the RD–RO relation in the writings of Barry Barnes. As will be seen, Barnes' position incorporates both realist and constructivist themes. Particularly problematic in Barnes' account is his use of a pictorial model of representation regarding the RD–RO relation. The issue is whether such a model is consistent with a constructivist programme.

My thesis throughout is that to the extent the realist-constructivist debate regarding representation in science is posed as an either–or situation between mutually exclusive and contrary options, the debate is a red herring. A few recent writers appear to recognize this; too many do not. For this writer, *the* salient issue is the extent to which realist and constructivist elements are mutually at work and interactive in the design and utilization of RDs in scientific contexts.

2. One constructivist account of representation

In *Interests and the Growth of Knowledge* (1977), Barnes examines various instances of RDs, with emphasis on the socially-constructed and socially-defined relation between these RDs and a given represented object or RO. Barnes' examples of RDs include: diagrams, a city map, a medical illustration of the musculature of the human arm, and contour lines on an elevation map. The issue Barnes addresses concerns the socially constructed and defined representational accuracy between RD and RO and the extent to which criteria of accuracy are socially contingent. In addition, for Barnes, considerations of accuracy are not independent of users' "procedures, competencies, techniques and objectives." As Barnes (1977:6) remarks,

> Representations are actively manufactured renderings of their referents, produced from available cultural resources. The particular forms of construction adopted reflect the predictive or other technical cognitive functions the representation is required to perform when procedures are carried out, competencies executed, or techniques applied. Why such functions are initially required of the representation is generally intelligible,

> directly or indirectly, in terms of the objectives of some social group.

I fully concur with Barnes' account of symbolic representations (wiring diagrams, maps, etc.) as 'analogous to techniques' rather than as merely objects of passive contemplation. Any account which divorces RDs from the contexts of *praxis* that define and concretely situate such devices clearly ignores a salient – perhaps *the* salient – influence on the construction and utility of RDs. (For further discussion of this point I highly recommend Hacking's *Representing and Intervening*, 1983; esp. 130–146.)

Unfortunately, the examples of ROs Barnes specifically employs are physically static and spatially fixed: circuit boards, built environments, musculature and elevations are non-dynamic structures. Barnes apparently takes such structures and their RDs as paradigmatic of the representational relation. For example, terrain maps – surely a case of a static RD of a static RO – constitute for Barnes (1977:7), "one of the clearest and most accessible contexts in which to examine the connection between the structure of representations and their function." Consequently, the static examples of RDs Barnes cites by no means exhaust the range of such devices in everyday or scientific inquiry. As suggested above, Barnes' examples do not extend to: (i) ROs in a dynamic, non-static state or (ii) temporally-changing events and processes. Given (i) and (ii), the static character of maps, schematic diagrams and textbook illustrations can not in principle adequately represent change – except, of course, through a series of such diagrams or illustrations. For example, the terrain effects of erosion, aircraft performance or urban expansion can be expressed by a series of maps at different time frames, but the dynamics and continuity of such changes are not captured. Even more problematic for Barnes' account is when there are no spatially-defined *objects* to represent (e.g., airfoil lift-drag coefficients, metereological phenomena, unemployment rates or demographic changes).

Perhaps *the* problem with Barnes' analysis of RDs is that it unnecessarily restricts itself to *a picturing model of the RD–RO relation*. As Barnes (1977:4) remarks,

> Our strategy should be to reveal pictorial representations ... as typical of representation and knowledge generally.

Again (1977:7), as pictorial representations,

> Maps indeed afford one of the clearest and most accessible contexts in which to examine the connection between the structure of representations and their function.

In many of the examples Barnes employs, the RD maps onto *structural features* of the RO. Such mapping relations are the easiest representational relations to deal with, so it is understandable why Barnes focused on *these* particular relations. However, the mapping model does not work where spatial and contour features of the RO are either absent or irrelevant. For instance, an EKG record does not in any literal/representational sense provide a topographical-like map of the electrical activity of the heart in the way that Barnes' wiring diagram does of an electrical circuit. Nor does a chemical formula map onto molecular structure in any obvious topological fashion! Given these counter-examples it is far from clear that pictorial representation is "typical of representation and knowledge generally," or that maps constitute a paradigm case of RDs.

Admittedly, some RDs do map onto their referents in a fashion. For example, the RDs (arrows, curving lines, numbers, etc.) used on weather maps do depict highs and lows, cold and warm fronts, wind direction, pressures, etc. But it should not be concluded that this constituted a one-to-one mapping between RDs and real (weather) events. On the contrary, such RDs are not givens but constructs. The history of measuring, encoding and depicting weather phenomena shows that meteorologists' RDs are but shorthand devices for representing barometric, temperature and wind speed/direction measurements. Clearly the concept of an isobar (a line connecting equal pressures), rather than literally representing something, was simply borrowed from the fiction of contour lines (equal elevation) on geological maps. What then do isobars (or contours lines) map onto? Nothing. On a constructivist account they are simply useful conceptual devices for introducing order into measurements and, in turn, for extrapolating from past to future data points.

Consequently, Barnes' account in *Interests and the Growth of Knowledge* deals with a relatively narrow and non-representative range of RDs, namely, those purportedly mapping onto structural features of spatially and temporally static objects. Furthermore, *in contrast with questions of utility, the fidelity with which RDs picture their corresponding ROs may be the least interesting issue to raise regarding scientific representation.*

In any case, and in spite of the serious conceptual problems noted in the preceding paragraph, Barnes' account of representation clearly endorses a constructivist version of the creation and deployment of RDs. This follows given that for Barnes all RDs are "actively constructed assemblages of conventions or meaningful cultural resources ..." (1977:9). Again, the significance of RDs – and of scientific knowledge in general – necessarily requires reference for Barnes to '*institutionalized* technical procedures' (1977:9; emphasis added).

3. The realist position regarding represented objects

As to the nature of the RO, Barnes never seriously doubts that there is a representation-independent reality which ultimately grounds RDs. The reason why scientific knowledge – and empirical knowledge in general – is useful "is precisely because *the world is as it is*" (1977:10; emphasis added). Consequently, knowledge is "a function of what is real, and not the pure product of thought and imagination." Elsewhere, Barnes (1977:25–26) adds that there is a single reality which is mind and culture independent and which can not be symbolically captured in any finite set of RDs.

> Everything of naturalistic significance would indicate that there is indeed one world, one reality, 'out there', the source of our perceptions if not their total determinant ... Reality is the source of *primitive causes*, which, having been pre-processed by our perceptual apparatus, produce changes in our knowledge and the verbal representations of it which we possess.

Bhaskar (1981:363) also claims that the extent to which scientific theories are cumulative and amenable to revision,

> strongly indicates such theories are (fallible) attempts to describe real states and structures of Nature, as they succeed one another in providing better accounts of a mind-independent reality.

In a similar realist vein, Bloor (1976:36–37) presupposes the existence of a common external world 'with a determinate structure.' This external world

> is assumed to be the cause of our experience, and the common reference of our discourse ... Often when we use the word 'truth' we mean just this: *how the world stands* ... In particular the assumption of a material world with which men establish a variety of different adaptations is exactly the picture presupposed by the pragmatic and instrumental notion of correspondence ...

Obviously, neither Barnes, Bhaskar nor Bloor would subscribe to that (extreme) constructivist account which argues that the so-called 'mind- and culture-independent reality' of the realists (particularly re. micro-level phenomena) is: (i) a mere conceptual posit, an extrapolation from theory-interpreted data points, and therefore (ii) a digression (or even regression) into speculative metaphysics. Either we remain silent about this supposed independent reality (which is Wittgenstein's proposal in the last line of the *Tractatus*: 'Whereof one cannot speak one must remain silent'), or we proceed to describe in some detail just what this reality consists of. The problem, though, with this latter tack is that even if we employ our 'best-informed and current scientific knowledge' our descriptive accounts of reality will be contingent on the explanatory Kuhnian-like paradigms currently in vogue. Accordingly, what inquirers are prepared to posit as a real entity or causal relation today may be demoted to artefactual status tomorrow. So, when is this inventory of reality to be taken? Certainly not today; perhaps only in Peirce's 'long run.' (As Peirce, 1885:8.41, once remarked: "the real is that which any man would believe in, and be ready to act upon, if his investigations were to be pushed sufficiently far.")

This reply does not of course conclusively prove that the realism of Barnes, Bloor or Bhaskar is conceptually untenable. Still, consideration of the interpretive and constructive dimension in scientific inquiry makes one hesitant to endorse any ontology which grounds itself in 'the best available scientific evidence.' Such a realist reconstruction also presumes that scientific knowledge is cumulative, progressive and approximates to reality over time. *Obviously, such an assumption begs-the-question concerning an independent reality*! Finally, such a realist ontology could also be said to be irrelevant and empty baggage *given that it tells us nothing whatsoever regarding the day-to-day interpretive, social and negotiative activities associated with the generation and legitimation of RDs and ROs.*

In their account of knowledge construction and legitimation in science, realists consistently ignore such inquirer-contingent considerations as largely irrelevant to epistemological matters proper. Unfortunately for this line of thinking, to understand the conceptual career of RDs and ROs no sharp demarcation can in fact be drawn between the (sociological) context of discovery and the (epistemological) context of justification. Woolgar (1983:246) provides an account of how realists would submerge the interpretive/constructivist dimension in inquiry:

> an initial use of accounting procedures to *constitute a new reality is subsequently regarded as no more than an attempt to* report upon or *reflect* what was there all along ... [Accordingly,] the reversal of the connection between account and object also entails the removal from the scheme of any constitutive activity of the discoverer.

In *Art and Artifact in Laboratory Science*, a recent study of scientists' situated laboratory practices and the achievement of order, Lynch (1985:202–216) concluded that: (i) inquirers' technical talk and technical work serve to establish and define objectivity, and apart from such considerations no independent *a priori* standard of objectivity would be available to inquirers in evaluating 'real-world objects,' (ii) in opposition to a realist account, the object of reference (the RO) in inquirers' situated discourse is not an integral phenomenon with a unified meaning, (iii) rather than 'objectivity' being lost in the dialectical exchange between inquirers' differing accounts, such disagreement and eventual convergence serves to establish and define objectivity, (iv) the objectivity of the RO is therefore constructed over time rather than an ontological given and, consequently, (v) scientific accounts re. the objectivity of the RO only "gradually emerge within a collaborative setting of inquiry" (1985:215).

In *The Manufacture of Knowledge*, a similar argument is advanced by Knorr-Cetina (1981:3):

> Rather than considering scientific products as somehow capturing what is, we will consider them as selectively carved out, transformed and constructed from whatever is. And rather than examine the external relations between science and the "nature" we are told it describes, we will look at those internal affairs of scientific enterprise which we take to be *constructive*.
>
> [On a realist account,] we tend to think of scientific "facts" as given entities, and not as fabrications. In the present study, the problem of facticity is relocated and seen as a problem of (laboratory) fabrication.

Regarding the supposed theory- and interpretation-neutral character of data points, and the blurred distinction between fact and artefact, the following discussion is relevant. In the late 1960s, Jocelyn Bell, a research assistant in the radio astronomy laboratory at Cambridge, observed "the persistent appearance of a strange section of 'scruff'" on the recordings from equipment searching for quasars. The question was whether to interpret this scruff as fact or as instrument-generated artefact. For Latour and Woolgar (1986:33) what is particularly relevant to a constructivist reading of science is

> the method by which Bell made sense of a series of figures such that she could produce the account: "There was a recurrence of a bit of scruff." The processes which inform the initial perception can be dealt with psychologically. However, our interest would be with the use of *socially available procedures*

> *for constructing an ordered account out of the apparent chaos of available perceptions.* (Emphasis added.)

In addition, as the explanatory paradigms evolve so will descriptive accounts of what constitutes 'reality.' For example, RDs of the neural circuitry of the primate and human brain of the earth's weather systems, of the properties of pions, sister and daughter muons and neutrinos, etc., are contingent on existing theory, paradigmatic assumptions, technical recording apparatus (Latour and Woolgar's, 1986:51, 'inscription devices'), and so forth. The reality descriptions science generates are as much a function of these considerations as of Barnes' (1977:25) so-called "one world, one reality, 'out there,' ... the source of *primitive causes*."

Regarding such contingencies of theory, interpretation and apparatus on what we take to constitute reality and facticity, Knorr-Cetina (1981:33) seriously questions whether

> the problem of facticity is to be located in the correspondence between the [cognitive] products of science and the external world ... The process of scientific enquiry ignored by objectivism (its "context of discovery") is itself the system of reference which makes the objectification of reality possible ... Thus, the problem of facticity is as much a problem of the constitution of the world through the logic of scientific procedure as it is one of explanation and validation.

Consequently, Knorr-Cetina rejects any neat separation between the context of discovery and the context of explanation. Nor, as claimed earlier, is there a clear demarcation between, on the one hand, the reality-denoting element in RDs and, on the other, the conventional, constructivist and theory-contingent dimensions to scientific representations. Ironically, while clearly recognizing the constructivist component to RDs in science, Barnes would bifurcate these RDs into: (i) their conventional/constructivist dimension as against (ii) their reality-denoting feature. Once we recognize (i) then reference to (ii) can constitute nothing more than a Kierkegaardian 'leap of (realist) faith.' It is not accidental that Barnes simply packs his claims re. (ii) into a few brief and unargued-for propositions. Once granting the importance of (i), why does Barnes – or any other constructivist sociologist of science for that matter – feel it necessary to go on and make metaphysical claims re. the 'one reality out there,' the supposed ultimate primitive cause of our empirical knowledge? Why not simply concentrate on the conventional/constructivist dimension to RDs and scientific knowledge and leave the 'deeper' and, in my opinion, epistemologically unresolvable issues, to the philosophers?

There is the possible rejoinder to the stand taken here that any account of RDs in science which omits reference to what RDs ultimately represent would render totally unintelligible why inquirers find useful some RDs over others. Continuing the rejoinder, *the* reason why at least some RDs (e.g., the three-dimensional DNA model of Watson, Crick and Watkins or Pauli's model of the neutrino and its properties) function as such powerful predictive and explanatory tools in science is simply because such models more and more come to approximate reality through successive model revisions. This is a strictly pragmatic appeal to the heuristic value and utility of (at least some) RDs. The logic of the argument is impeccable if one assumes the initial premise that a RD could have utility value if and only if it accurately or truthfully (another favorite realist expression) represented what it portends to represent, namely, reality or at least some segment of reality.

But is this necessarily the case? The Ptolemaic, geocentric system of stellar, lunar and planetary representations is generally considered as inaccurate. Still, it was employed until the advent of the Loran system and satellites to constitute the basis for both maritime and aerial navigation – with surprising predictive accuracy and theoretical coherence. Dalton's and, later, Rutherford's model of the atom is another example. Both these macro- and micro-level models now belong to the history books. (For an interesting discussion of the cognitive maps and RDs employed by the ancient sea-faring Polynesians, see Oatley, 1977.)

An RD can therefore possess heuristic value without necessarily mapping in any truth-preserving fashion onto anything at all. That is, the RD need not preserve or capture anything about a so-called mind- and culture-independent reality. I do not have a ready reply to the realist's question, 'Why do some RDs have more heuristic value than others?' I am not questioning at this point that some RDs do indeed have such value. However, I do not feel that the issue is resolved by such simplistic answers as, 'Well, it's their approximation to reality that grounds their explanatory and predictive utility!' *The* problem here is the term 'approximation.' Once we seriously question the pictorial metaphor re. the RD–RO relation, then have we said anything at all with such locutions as, 'approximation' (correspondence) to reality? Such locutions, and the language-game in which they are embedded, continue to draw on a simplistic picturing theory of representation.

4. Lessons to be learned

Such discussions and controversies continue to divert our attention from the interpretive factors in scientist's construction and reconstruction of RDs over time. Additionally, certain recent case studies in this area (e.g., Collins, 1981b, 1985; Garfinkel, Lynch and Livingston, 1981; Knorr, Krohn and Whitley, 1981; Knorr-Cetina, 1981, Knorr-Cetina and Mulkay, 1983; Latour and Woolgar, 1986; and Lynch, 1985) show that inquirers' accounts of the supposed mind- and culture-independent reality of Barnes is certainly subject to conceptual revision over time. Obviously, for realists ontological assumptions constitute a *sine qua non* for understanding scientists' empirical and theoretical activities and apart from such a realist ontology such activities would make no sense at all. For constructivists, on the other hand, *the* issue is not the respective merits of one ontological framework over another but inquirers' discourse, accounts, cognitive constructions and *praxis*. Besides, as mentioned above, a serious problem with any realist ontology is that any set of reality-denoting claims to which inquirers are collectively prepared to attach evidential and existential import varies over time and is continually subject to negotiation and revision. *Accordingly, it is simply inconsistent with any thorough-going constructivist strategy to venture into realist – or perhaps even anti-realist – claims.*

At least one philosopher of science has recently argued that what we take to constitute 'reality' changes with our representations and RDs. Accordingly, on this account no neat, simplistic distinction is possible between Nature and (representational) Convention. Hacking (1984:139) concluded from the suggestions of Kuhn and others that

> with the growth of knowledge we may, from revolution to revolution, come to inhabit different worlds. New theories are new representations. They represent in different ways and so there are new kinds of reality. So much is simply a consequence of my account of reality as an attribute of representation.

Continuing, Hacking (1983:136) added that,

> It will be protested that reality, or the world, was there before any representation or human language. Of course. But conceptualizing it as reality is secondary. First there is the human thing, the making of representations. Then there was the judging of representations as real or unreal, true or false, faithful or unfaithful. Finally comes the world, not first but second, third or fourth.

One does not have to endorse all the claims of early-twentieth century positivism to see the merits of avoiding the murky claims and even murkier arguments of the realist/anti-realist debate. Hacking (1983:145) similarly concludes that realism and anti-realism "scurry about, trying to latch on to something in the nature of representation that will vanquish the other. There is nothing there." I for one believe that a thorough-going constructivist account of representation and RDs in science *is* possible which avoids any entanglement with this unresolvable and debilitating realist/anti-realist debate. Such an account would look quite different from that proposed, for example, by Barnes. (For a pragmatic statement of what such a disentangling strategy would look like, see Collins, 1981a, 1983a, for discussion of both the empirical and the radical programmes of relativism. For a defense of the latter programme, see Tibbetts, 1985, 1986, 1987.)

For one thing, it is clear that a constructivist/relativist account would not be committed – as is Barnes – to the pictorial account of representation discussed earlier. The pictorial metaphor for Barnes presumes some sort of one-on-one mapping relation between RD and RO. With the emphasis now shifted to the RD and its contingent construction and revisions over time, and away from Barnes' 'independently-existing RO,' the language-game of mapping relations loses its logical force. For realists the pictorial model re. the RD–RO relation will still constitute a useful explanatory metaphor. However, keeping in mind my earlier criticisms (that the mapping model of Barnes is unable to express spatially- and temporally-dynamic properties), I would contend that even as a metaphor the language-game of mapping relations is seriously deficient. Additionally, there will be no talk about the extent to which RDs (whether two-, three- or n-dimensional models, formulae, graphs, etc.) preserve or capture the so-called essential (structural) features of what is represented. *What* it is that is ultimately represented by RDs and with what degree of accuracy and fidelity will be of no concern for this proposed thorough-going constructivism.

I therefore conclude that there is no single nor fixed answer to the question, 'What after all do RDs in science represent?' Nor is the looked-for-answer located outside the domain of scientists' constructivist and interpretive activities concerning both *what* is being represented and *how* it is best represented. The realist's purported distinction between Nature and Convention can only be sustained by ignoring the hermeneutics of scientific representation.

References

Barnes, B. (1977). *Interests and the growth of knowledge*. Boston: Routledge and Kegan Paul.

Barnes, B., and Edge, D., Eds. (1982). *Science in context: Readings in the sociology of science.* Cambridge: MIT Press.

Bhaskar, R. (1978). *A realist theory of science*. New York: Humanities Press.

Bhaskar, R. (1981). Realism. In W. Bynum et al. (Eds.), *Dictionary of the history of science*, 362–363. Princeton: Princeton University Press.

Bloor, D. (1976). *Knowledge and social imagery.* Boston: Routledge and Kegan Paul.

Bohme, G. (1977). Cognitive norms, knowledge interests and the constitution of the scientific object. In E. Mendelsohn, P. Weingart and R. Whitley (Eds.), *The social production of scientific knowledge*, 129–141. Dordrecht: Reidel.

Collins, H. (1981a). What is TRASP?. The radical programme as a methodological imperative. *Philosophy of the Social Sciences* 11(2):215–224.

Collins, H., Ed. (1981b). Knowledge and controversy: Studies of modern natural science. Special issue of *Social Studies of Science* 11:1–158.

Collins, H. (1983a). An empirical relativist programme in the sociology of scientific knowledge. In K. Knorr-Cetina and M. Mulkay (Eds.), *Science observed: Perspectives on the social study of science*, 85–114. London: Sage.

Collins, H. (1983b). The sociology of scientific knowledge: Studies of contemporary science. *Annual Review of Sociology* 9:265–285.

Collins, H. (1985). *Changing order.* London: Sage.

Collins, H., and Pinch, T. (1982). *Frames of meaning.* Boston: Routledge and Kegan Paul.

Garfinkel, H., Lynch, M., and Livingston, E. (1981). The work of a discovering science construed with materials from the optically-discovered pulsar. *Philosophy of the Social Sciences* 11(2):131–158.

Grenier, M. (1983). Cognition and social construction in laboratory science. *Society for Social Studies of Science Review* 1(3):2–16.

Hacking, I. (1983). *Representing and intervening.* New York: Cambridge University Press.

Hollis, M. (1982). The social destruction of reality. In M. Hollis and S. Lukes (Eds.), *Rationality and relativism*, 67–86. Cambridge: MIT Press.

Hollis, M., and Lukes, S., Eds. (1982). *Rationality and relativism.* Cambridge: MIT Press.

Jarvie, I. (1983). Rationality and relativism. *British Journal of Sociology* 34(1):44–60.

Jarvie, I. (1984). *Rationality and relativism.* Boston: Routledge and Kegan Paul.

Knorr, K., Krohn, R., and Whitley, R., Eds. (1981). *The social process of scientific investigation.* Dordrecht: Reidel.

Knorr-Cetina, K. (1981). *The manufacture of knowledge: An essay on the constructivist and contextual model of science.* New York: Pergamon.

Knorr-Cetina, K. (1983). The ethnographic study of scientific work: Towards a constructivist interpretation of science. In K. Knorr-Cetina and M. Mulkay (Eds.), *Science observed: Perspectives on the social study of science*, 115–140. London: Sage.

Knorr-Cetina, K., and Mulkay, M., Eds. (1983). *Science observed: Perspectives on the social study of science.* London: Sage.

Latour, B., and Woolgar, S. (1986). *Laboratory life*, 2nd ed. Princeton: Princeton University Press.

Laudan, L. (1977). *Progress and its problems.* Berkeley: University of California Press.

Laudan, L. (1981). The pseudo-science of science? *Philosophy of the Social Sciences* 11(2):173–198.

Lynch, M. (1985). *Art and artifact in laboratory science.* Boston: Routledge and Kegan Paul.

Mendelsohn, E., Weingart, P., and Whitely, R., Eds. (1977). *The social production of scientific knowledge*. Dordrecht: Reidel.

Oatley, K. (1977). Inference, navigation and cognitive maps. In P.N. Johnson-Laird and P.C. Wason (Eds.), *Thinking: Readings in cognitive science*, 537–547. New York: Cambridge University Press.

Peirce, C. [1878] (1931–35). *Collected papers*. Cambridge: Harvard University Press.

Pickering, A. (1984). *Constructing quarks*. Chicago: University of Chicago Press.

Popper, K. (1972). *Objective knowledge*. New York: Oxford University Press.

Tibbetts, P. (1985). In defense of relativism and the strong programme. *British Journal of Sociology* 36(3):471–476.

Tibbetts, P. (1986). The sociology of scientific knowledge: The constructivist thesis and relativism. *Philosophy of the Social Sciences* 16(1):39–57.

Tibbetts, P. (1987). *The sociology of scientific knowledge: Problems of knowledge-generation and -legitimation*. Unpublished manuscript.

Woolgar, S. (1983). Irony in the social study of science. In K. Knorr-Cetina and M. Mulkay (Eds.), *Science observed: Perspectives on the social study of science*, 239–266. London: Sage.

The fixation of (visual) evidence

K. AMANN
K. KNORR CETINA *
Center for Science Studies, University of Bielefeld, P.O. Box 8640, D–4800 Bielefeld, FRG

1. Introduction

The fixation of belief, or consensus formation in science as sociologists are wont to call it, refers to a process whereby theories or theoretical hypotheses come to be accepted as fact in a community of specialists. In this paper, we shall be concerned with the fixation of "evidence" or of "sense data", a slice in the process of fact construction:

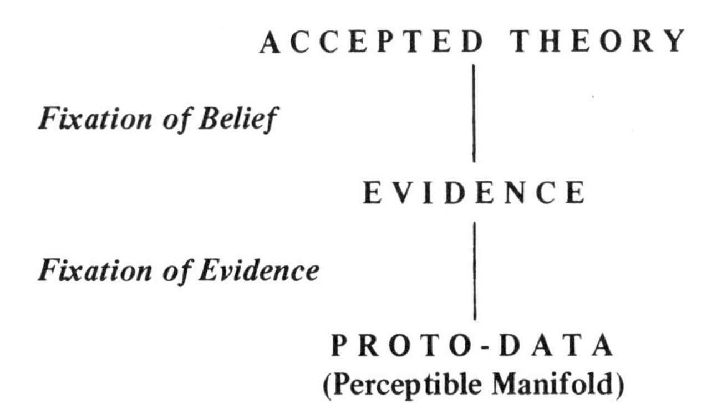

According to the standard view of science (Mulkay, 1979: Ch. 1), sense data are what we obtain when we test theories through experiments, and sense data tell us whether a particular theory is likely to be correct. This view has been undermined by the Du-

* We are grateful to Michael Lynch for his very helpful comments and suggestions.

hem-Quine thesis of underdetermination, according to which data can never conclusively prove or disprove a particular theory.[1] And it has been challenged by the claim that what counts as appropriate evidence in a theoretical controversy is itself negotiated during the controversy, hence evidence cannot serve as an independent arbiter of scientific belief (Collins, 1975). However, these challenges of the standard view deal only with the degree to which evidence is, from a logical point of view, pertinent to theory choice. They are not concerned with how sense data may be problematic in ways other than in relation to what they achieve in theory debates.

But are sense data problematic? Consider that in the natural sciences evidence appears to be embodied in visibility; in a literal sense, it is embodied in what we can see on a data display. Thus understood, the notion of evidence is built upon the difference between what one can see and what one may think, or have heard, or believe. Among these modes of relating to an object, only seeing bestows on objects an accent of truth. But does it really? And can we consider seeing as a primitive (in the sense of unconstructed), "truth-transporting" activity? We know of course that processes of seeing are subject to cultural and historical conventions, and that what participants see may depend on the institution of seeing involved (Gombrich, 1960). In regard to science, Kuhn (1970) has argued that consensual ways of seeing are maintained through shared paradigms, consisting of rules and standards for correct scientific practice. Under this view, what scientists observe should be grounded in their complex commitments to particular research traditions. Yet in the science we study, the problem appears not to be, as Merleau-Ponty said (1962:78), that "what you see depends on where you sit", but rather "nothing is more difficult than to know exactly just what we do see". Whatever role perceptual grammars may have in shaping what counts as evidence in disciplinary traditions, these grammars do not resolve the manifold problems associated with visual sense data in day-to-day laboratory work. The point is that just as scientific facts are the end product of complex processes of belief fixation, so visual "sense data" – just what it is scientists see when they look at the outcome of an experiment – are the end product of socially organized procedures of evidence fixation.

When we mention "seeing" in this context, we do not just mean sensory activation by some perceptible manifold-out-there. Most arguments which relate to "seeing", for example Quine's point about the equivocality of ostension in identifying visual objects (1960) and Campbell's attempted rebuttal (1986), presume a relationship between "seeing" and the linguistic reference to objects: "to see" an object is to recognize and at the same time to linguistically identify an object. But, what if these objects are, as they appear to be in science, *visually flexible phenomena* whose boundaries, extension and identifying details are themselves at stake? The problem for scientists is not the equivocality of ostension or the impossibility of being certain that a "translation" into language is correct. Instead, for practicing scientists, the difficulty of coming up with a translation in the first place is the prime concern. When we refer to processes of evidence-fixation, we refer to processes of developing and solidifying such translations.

In this paper, we offer an initial description of the kinds of mechanisms and processes involved in evidence-fixation. The data presented derive from an ongoing laboratory study[2] of molecular genetics conducted since September 1984 at the Center for Molecular Genetics, Heidelberg, FRG. The group studied works on transcriptional control mechanisms, that is on DNA regulatory elements which can dramatically increase transcriptional activity during the transcription of DNA into RNA and which, for that reason, are relevant to the understanding of normal and abnormal cell growth. The group publishes regularly in journals such as *Nature and Science*; it is one of the leading research units in the area on a worldwide basis. The leader of the group, who is also a Professor at the University of Heidelberg, and its core-members spent several years in the United States, and two American post docs were employed in the unit during the period of observation. The Center is basically financed by government sources; the research is done by post docs, doctoral students, and students working toward the equivalent of an M.A. Most of the examples presented in this paper derive from a series of interconnected experiments involving a particular method of RNA preparation ("Sl analysis").

2. Sense-data and evidence

To begin with, we will introduce a distinction between the "data" recognized in the laboratory and the "evidence" published in scientific papers.[3] In the molecular genetics lab we describe, there are at least three different modes of practice through which materials in the laboratory are visually inspected, and through which seeing becomes a distinct, specially marked activity in the stream of laboratory shop work:

1) The first mode of practice involves techniques of manual and instrumental *enhancement*, such as in simple cases, holding a test tube against the light to assess the progress of a biochemical reaction, or taking a polaroid photograph (which participants call "fast picture") of an electrophoresis gel to check on the position of DNA fragments or the success of a plasmid construction.
2) The second occasion for visual inspection centers around "*data*" – which in the study of transcriptional control mechanisms and many other molecular genetics fields are mostly visual traces generated by radioactively marked DNA or RNA fragments separated in an electrophoresis gel on which an X-ray film has been exposed. The following exhibit (see Exhibit 1) offers an example of an autoradiograph film as it appears in the laboratory.[4]
3) The third set of practices revolves around "*evidence*", by which we mean the data actually included in scientific papers or shown in oral presentations. Data become evidence only after they have undergone elaborate processes of selection and transformation.

Now seeing becomes problematic only in the second case, when scientists deal with "data". The distinguishing characteristic of the first set of practices is that they tinker with the conditions that improve the visibility of certain materials. But the visual materials themselves appear unproblematically readable, and the pictures created on this level have only *local* relevance. They are not normally discussed at length among participants or displayed in the papers produced. Any problems with "enhance-

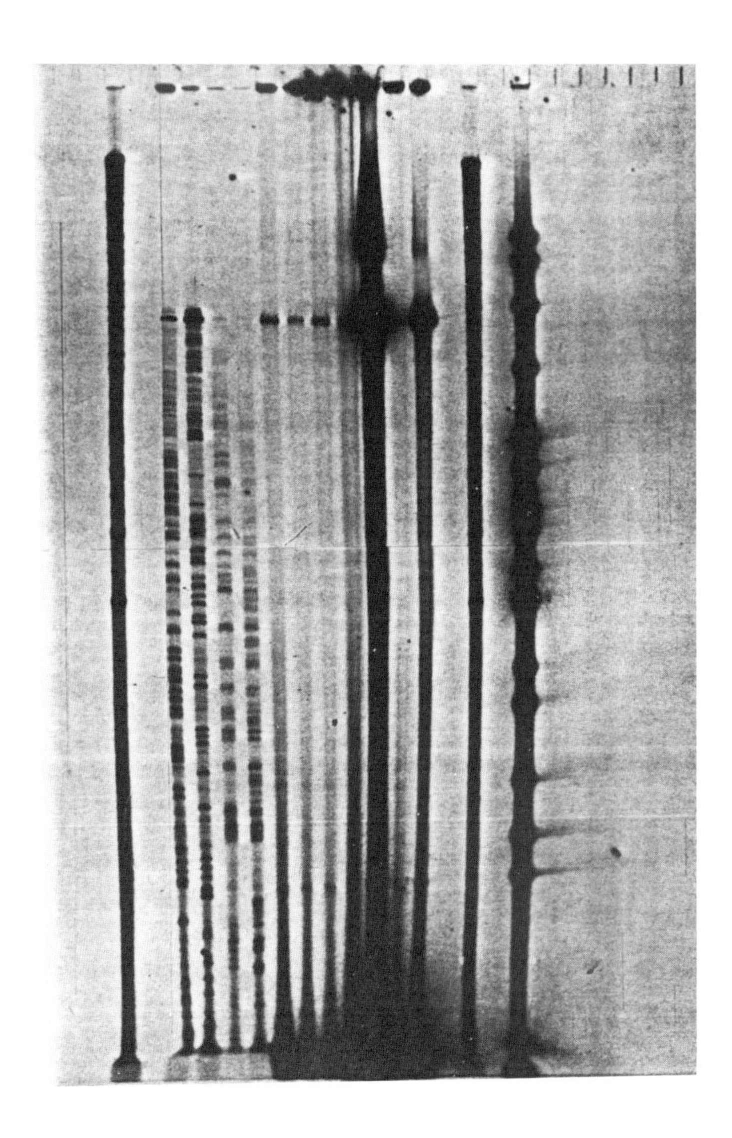

Exhibit 1. Example of an (unedited) autoradiograph film as it appears in the laboratory

ments" tend to get quickly resolved, and they are treated as significant only in regard to the control function they perform.[5] The instrumental techniques involved have sunk into the background of taken for granted devices in the pursuit of other, more "interesting" matters. Not so with "data". The autoradiograph data which are the focus of the second set of practices lie at the very center of scientists' attention, and they form the core of the papers produced. The distinguishing characteristic of visual data is that they are not, like the "enhancements" mentioned before, treated as unproblematic displays of visual objects. Data act as a *basis for sequences of practice* rather than observation at a glance. They are subjected to extensive visual exegeses, rendering practices which attempt to achieve *the work of seeing what the data consist of.* The question of interest to the analyst in these visual exegeses is "what do we see". The image, here, becomes a "workplace" (Lynch, 1985b) for participants in seeking an answer to this question. The sociologically interesting phenomenon is that *seeing is work*. But what sort of work?

3. The machinery of seeing

Characteristically, autoradiograph displays appear in the laboratory when an author retrieves them from the film room where they were exposed for a number of hours or days, and starts to inspect them against the light. (See Exhibit 2)

Other researchers present in the laboratory are attracted by such events, gather around the visual materials, finger the documents and gaze about their surfaces. As they examine the film, scientists begin a series of verbal exchanges. This is where language becomes relevant in the present context. But note: the resulting perceptual identification is not just the product of *language*, it is the product of conversational *talk*. What difference does this make? When embedded in talk, "seeing" is interactively accomplished. Thus the process is not just a semiotic process, in the sense of involving a translation into a generalized system of signs. Nor is it mainly a cognitive or interpretative process in the sense of involving individual conceptual decoding. Instead, the process has a *speech act* and particularly a *dialogical* or *interactive* structure.[6]

Exhibit 2. Participants looking at an autoradiograph film against the light

Thus, when practitioners encounter the "external world" in terms of the sense data described, the *machinery of seeing is talk*. This talk has several specific characteristics: First, image analyzing talk is *attached to objects*, specifically to the data displays ("films") which are the subject of the exchanges. Image analyzing exchanges are not just "about" an object; they are also "with" an object. One might say that participants interact not only with each other but also with the object to which they attach their comments. Significantly, the objects addressed by participants are also manipulated during these exchanges. The operation performed, the detail observed, complement the utterances as concrete but non-verbal "phrases". Furthermore, the talk produced appears to some degree to be *organized by* the documents inspected. This *documentary organization of talk* is also found when scientists discuss the content of the papers they are writing, as presumably it is whenever talk is concretely related to objects which exhibit their own semiotic organization.

A second characteristic of image analyzing talk is that it is embedded in a *series of exchanges* which are interconnected by one or several related displays. Related displays are displays which derive from replications or slight variations of the same experimental procedure. Participants tend to return repeatedly to the same or related displays to discuss their content, a feature of shop exchanges made possible through the continued accessibility of participants to each other while they work in the lab. Participants may vary in these exchanges: *it is the image which integrates the series*, not the continuity of speakers. Serial exchanges of this sort indicate practitioners' *occasional presence* to complex situations and the *local* and transient character of most problem solutions. Participants do not seem to resolve issues raised by the features of an image once and for all. Instead, they repeatedly "visit" a problem, thus continually reopening cases that, as judged by an outside observer, seem to have been closed by a definitive conclusion the last time the problem was considered.[7]

There are also more general characteristics of shop talk which might be noted, for example the phenomenon indicated before that the substance of the talk is interactively or collaboratively produced. Speakers' contributions remain oriented to each other within conversational turns of roughly equivalent length, such that the substance of the talk is the outcome of joint conversational work. Ostensibly, what is achieved in these transactions is technical work and not, to borrow a distinction by Goffman (1971: 147–148) ritual or relationship work (though the latter may be performed through the former). What is the pattern of interactional organization in these exchanges? We want to offer some observations on the interactional shape of film talk and on the conversational devices participants employ in performing image analysis work.

4. The interactional organization of image analyzing talk

First the pattern of interactional organization. When two or more participants gather around an autoradiograph display in the lab, they face the task of finding their way about the film – that is, of identifying various black and white bands on the film and the objects these bands represent (see Exhibit 1 above). In general, practitioners go about this task by asking a series of questions. These typically refer to where on the film are the following constituents:

- the "marker", a known construct usually inserted into the first and/or last lane of the electrophoresis gel. The marker supposedly yields a known pattern of bands which serves as a measuring stick for the length of the DNA and RNA fragments under investigation;
- the "probe", a radioactively labelled DNA fragment to which RNA is hybridized and which appears in all lanes in a specific position;
- the "starts", that is the expected bands which indicate the molecules separated in the gel run;
- the "length" of these items, that is the position of the bands on a vertical scale as determined by external reference tables that indicate the expected "length" of the marker bands;
- in addition, there occur opening questions which determine the general nature and identity of the film, the stage of the analysis, the display a film compares to, etc.

With different displays, different objects may become relevant, such as "windows" (white spots) on "footprints"; yet typically there are inquiry sequences through which practitioners attempt to specify the geography of the display. In exchanges between two or more persons, questions are always posed to the author of the film by a recipient as he or she seeks to learn more about the film. Students of institutional encounters such as medical interviews, calls to the police or classroom interaction have found that the person asking questions (the doctor, the police, or the teacher) appears thereby to dominate the encounter by placing limits on the placement and the content of recipients' responses (e.g., West 1983).[8] In contrast, the inquirers in film talk do not appear to exert, by adopting the role of the questioner, such power. At least on the face of it, the roles seem to be reversed: it is the person questioned who controls a valuable good, namely relevant information, whereas the questioner seeks to obtain a share in this good. Furthermore, there appears to be agreement among practitioners as to the questions which must be asked. Thus the questioner is not at liberty to shape the interaction by carefully choosing and editing his or her question. If there was no interaction, the author of the film would have to raise and answer the same questions, as ln. 323 of the following, monologic exchange indicates (the author, distracted by what he sees on film, apparently takes no notice of other parties' contributions). The exchange also illustrates the initial phase of an inquiry sequence: It begins by an inquirer (He) asking a general identity and recognition question (→ ln. 319) not answered by the author (Er) and continues by the questioner asking for the location of the marker (→ ln. 325) and the probe (→ ln. 326, 328). As his questions remain unanswered, the inquirer reverts to another sequence opener (→ ln. 329), i.e., to the question about the film to which the other film compares (about the experimental series to which the film belongs):[9]

(160102 85p98)

→ 319 He and, what is this?

320 Er ha, over night ((exposed)), exactly like last time. And what do you see ((holds up film))? Nothing! ((Pause))

→			Where is the probe anyway?
	324	Ni	simple enough, there is nothing on it
→	325	He	these are the markers, aren't they? Left and
→			right. This is the probe?
	327	Er	((remains silent))
→	328	He	this is the probe?
	329	Er	((remains silent))
→	330	He	which ((film)) does it compare to?
	331	Er	((annoyed)) what do you mean, which does it compare to?
		((Etc.))	

Note that inquiries into the geography of the film are sequentially structured in terms of a series of questions posed in a certain order (the identity recognition question is posed before the marker question, which in turn is posed before the probe-question; the question for the location of the starts comes last). Authors do not offer summary accounts of all the relevant information they possess in regard to the identity of the bands on film. This is one example of the more general phenomenon that complex problem situations appear to become *interactionally dissolved* in shop talk. In cases of image-analyzing talk, there is a perfect reason for this interactional dissolution. While authors of films have an informational advantage over non-authors, as acknowledged by their being consulted by the latter, they have few ready-made answers, and must find their way around the film at hand by inspecting the image just as non-authors must. If the questions indicated above could be readily answered, film analysis exchanges would presumably have a straightforward and readily intelligible structure including the following segments:

1. An *opening sequence* comprising a summons (such as a non-verbal display of the film which has the effect of a summons on participants within reach) and/or a verbal news announcement or news request followed by an answer.
2. An *information-gathering question-answer sequence* resulting in a specification of the geography (identity of the bands) and perhaps of the architecture (how the image was "built") of the film.

3. An *evaluative sequence* resulting in an evaluation of the expected bands, the actual results of the experiment.
4. A *resolve* or *performance recommendation* based upon the evaluation of the film which indicates the actions to be taken in subsequent experiments or in preparing the material for publication.

The whole exchange would have the character of a newsreport/newscommunication elicited by receivers, or of a collegial information-sharing encounter among fellow workers engaged in roughly similar tasks. However, we cannot offer an example of such an exchange. While fragments of the above structure can be found in all appropriate encounters, not one of the exchanges recorded is "*whole*" in the sense of displaying the projected structural form. The structure of actual film talk is characterized by the absence of a distinctly marked evaluation sequence and by a pattern of diversions from the remaining central piece, the inquiry sequence. The questions in this piece provide something of a skeleton which holds the conversation together; but they also serve as pegs on which a variety of other segments hang. Why these diversions? Because sooner or later, the author of the film appears to be unable to provide a satisfactory answer to the questions and the inquiry sequence gets stuck. Other conversational devices take over and propel the exchange for variable periods of time in a different direction. The side sequences[10] thus formed which break the inquiry sequence apart account for the "garland" structure of real time film talk. (See Diagram 1)

The left side of the diagram exhibits the projected path through an image analysis task as posed by autoradiograph displays in the lab observed, while the right side offers a schematic representation of the garland-structure of actual film talk. Projected paths are straightforward, recognizably rational, but nonetheless conventional[11] "throughways" through an image. However, we can only pursue them if nothing distracts our attention, and if there are no obstacles which force us to take a detour. Participants attempt to pursue projected paths; they continually initiate and return to the inquiry sequence in film analysis exchanges. Thus projected paths are also empirically recognizable, seemingly preferred forms of interactional organization and not merely sequences of steps

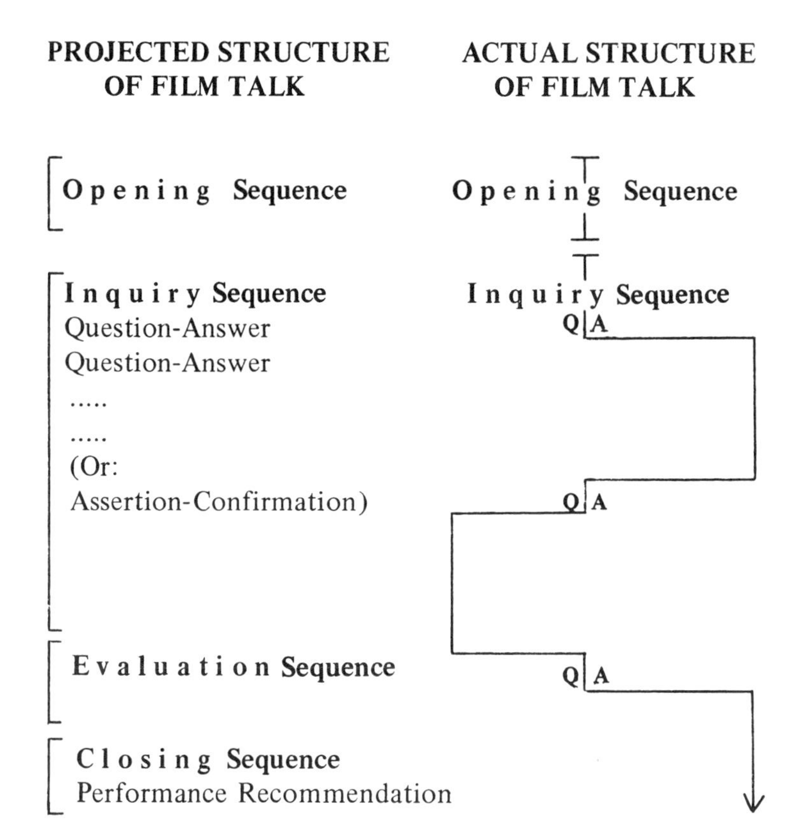

Diagram 1

that follow "logically" upon each other. Yet, for the reason indicated above, in real time passages through an autoradiograph film participants get caught up in side sequences, which make up the bulk of a series of exchanges.

What happens in these side sequences? Conversational devices other than the inquiry sequence take over the exchange. Three patterns of talk are particularly notable: in our understanding, these are general, inference producing devices employed in many problem situations.[12] Elsewhere we have called them procedural implicature, optical induction and the oppositive device (Amann and Knorr Cetina, 1988a). In the present case, participants "slide into" one or the other of these patterns as they run into problems with identifying and interpreting the bands on the film.

5. Conversational devices employed in image analyzing exchanges

5.1 Procedural implicature

Consider first the *procedural implicature* device.[13] This pattern is, in a sense, a variant of the inquiry sequence adapted to another use. It is employed to derive non-obvious conclusions from mute experimental outcomes by means of an inquiry into the procedures through which these outcomes have come about. In a nutshell, the exchange consists of a series of question-answer and/or assertion-confirmation adjacency pairs (pairs of utterances in which the first utterance constrains the second, as in a question which "demands" an answer) which access and make public indexical information from eyewitnesses of a phenomenon. As before, the author is not asked to provide a summary assessment of the situation. Rather, he or she is consulted in an iterative, stepwise fashion as a *living archive* of the details which constitute the time-space geography[14] of the film. The pattern may be followed by a conclusion in the form of an interpretation ("this means ...") or of a performance recommendation ("I would ...", "you've got to ..."), and it is frequently initiated by a statement which discloses some problematic occurrence or information.

The following example is the first of two interrogatory series (ln. 114–122 and 126–141) separated by a candidate interpretation (ln. 114). The series is part of an exchange which attempts to establish which of the bands on the film are correct "starts" and which are the "probe" or false starts. The procedural inquiry (→) is initiated after Ea indicates where approximately, in his opinion, the start side should be (ln. 110ff.)

(15018505ffp3)

	110	Ea	somewhere there. This is CAT and this has to be away. <u>There</u>! somewhat shor/ somewhat longer, merely 10 basepairs, right? More over <u>there</u>! This would be the level of
→	114	Jo	what probe is this?
	115	Ea	polyoma-CAT
→	116	Jo	I don't quite understand it yet. The/ this ha/ this was done with SV40, the polyoma?

	118	Ea	transfected. Transfected. And when I knock it down with my probe/
→	120	Jo	this is then knocked down with your probe?
	121	Ea	yes, and then I get only/ then I only get CAT protected ..
x	122	Jo	then this up to eco would have/
x	123	Ea	to be CAT

After two more turns in which Jo and Ea elaborate this interpretation, the inquiry pattern continues:

→	126	Jo	Could it be that you have a bad homology somewhere? Between your DNA and the probe?
	128	Ea	how do you mean
→	129	Jo	if, let's say you cloned in a way such that at the hindIII-cut...there was a missing homology ...I/ I don't know your clones/ if you'd/ 10 basepairs at the hindIII mark or s/ or 5 basepairs/
	134	Ea	but there is/ in there/ the hindIII side has been changed into a bglII side
	136	Jo	yes
	137	Ea	on this is cloned, polylinker. and into <u>that</u> I cloned
→	139	Jo	right and you took the <u>same</u> probe for the separation of strands?
	141	Ea	of course
	142	Jo	yes/ this can/ sure/ this can be through/ this can come about
x			if you have a missing homology at the
x	145	Ea	now listen, the
x	146	Jo	hindIII- side
x	147	Ea	problem is as follows ((Etc.))

Notice the overlaps (x) as conclusions are collaboratively produced (ln. 122ff.) or rejected (ln. 142ff.), and the fact that answers to a procedural question may take more than one turn (ln. 132f., 135). The latter switches the question-answer pattern to one of assertion and confirmation (ln. 132–134, 135–136 (first word only)).

5.2 Optical induction

Optical induction is a curious hybrid between visual operations and conversations.[15] With procedural implicatures, participants rely on the interrogation carried out to produce features of the history of a phenomenon which aid in object identification. With optical induction, in contrast, it is the image itself which prompts these features. In pursuing a procedural history, participants depart, for the time being, from the display which is the object of their talk. In performing optical inductions, participants concentrate on the image. Optical induction is a pattern which, for the most part, *consists of visual operations carried out through talk.* The linguistic means, however, are not question-answer adjacency pairs or assertions and confirmations, but sequences which include *formulations* of details of the bands on film mixed with interpretations. As participants inspect visual features of the film, as they pay attention to and compare the details of these features, they establish and reject candidate identifications of visual traces, and they do so in the sequential fashion typical of collaborative talk. The procedure is not linear, however. As participants move between traces in attempting to establish the identity of some bands by reference to others, they have frequent occasions to return to the same spot and to revise previous interpretations. Presumably, optical inductions occur in all image analyzing contexts, including those in other areas within and outside of science. The following example illustrates how participants' visual operations on the film (ln. 352–364) lead to certain interpretations (i.e., which band is the probe, that there is a transcript and where it is located; ln. 366–372), and how these interpretations in turn give rise to a new round of visual operations (ln. 371f.).

(1401188505ffp7)

352	Ea	(...) if you look at it ((points to film)) these run three times higher, okay? the difference here is a centimeter, this one, of this size
355	Jo	yesyes
356	Ea	okay. these up there don't even have half ((a centimeter)). You can do what you want, we just measured it. They're not on the same level.

The differences are very variable, depending, on the size of the fragment.

361 Jo if you shift this parallely with the others, right, like that, that way, that way, this is nonetheless not running on the same level as this

364 Ea no, this isn't on the same level, granted. I am not saying (it is), but I say/ this is/ say/

366 Mi but if these run on the same level, this greatly suggests, doesn't it, that this is the probe.

368 Ea sure this is the probe. But then I also know that I've got a transcript which runs all the way through

371 Jo but this can be this/ this one here. That is this band here.

((Etc.))

Which bands run "on the same level" is not obvious from what one sees on the display, as the transcript indicates. Participants must work out the looks and location of visual traces as much as they must work out which experimental variables the black and white spots on the film represent. How are they doing this? By shifting and otherwise manipulating visual traces (e.g., ln. 361) and by comparing the signs which appear in different locations (e.g., ln. 352). But also by taking some signs at face value for the time being ("if this is this then that is that..."), and by going back and forth between the geography and the architecture of the display (between the location of bands on film and the way the experiment was set up). As the above exchange continues, Ea counters Jo's proposal as to which bands might be the transcript/probe (ln. 371) by reference to the experimental design (ln. 373, 376):

371 Jo but this can be this/ this one here. That is this band here.

373 Ea but that's something else, wait, don't mix things up.

375 Mi yes but I mean, still/

376 Ea that's a different promoter, rightsh

In many respects, an autoradiograph film is like a maze designed by one participant in which the designer nonetheless finds him/herself lost. To locate a way out of the maze, participants identify and compare visual clues, point out where it might continue, follow some paths and recall the design of the maze to evaluate leads. Occasionally, they clash with each other about the direction to take. Then the pattern of talk becomes adversarial, and another routine of talk takes over the conversation.

5.3 The oppositive device

Possible turning points where the exchange could become adversarial are found in the above excerpt. For example, consider the counterproposal in ln. 371 or other contributions opening with a potentially oppositive "but...". Yet these possibilities are not taken up by participants as they work their way through the film in the above segment of talk. More suggestive examples of adversarial episodes are found elsewhere in the above exchange. To some extent, oppositive patterns of interaction "feed upon" or overlay other conversational patterns. This is because oppositive patterns are not only adversarial, they are also heavily argumentative; and participants may raise procedural questions or draw visual inferences in the service of their argument.[16] Oppositive patterns often start by one participant objecting against the proposal made by another. They continue by participants arguing with and negotiating about each others' candidate accounts of issues raised in the encounter. Like the other patterns mentioned, oppositive patterns occur in a variety of shop situations. In image analysis exchanges, oppositive episodes are likely to exhibit features of optical induction, as participants argue in terms of visual clues and operations. In the following exchange, which is part of the above series of conversations regarding the location of the probe and the starts on the film, such segments are found throughout the transcript. Consider the beginning (ln. 56–57) and then again ln. 63 and 67ff. of the transcript, in which the opponents compare their films and produce visual inferences (ln. 71ff.) while at the same time arguing about how preparatory conditions involving different salts influence the appearance of bands:

(1401198505ffp2)

56	Ea	if you want to say that you're seeing plus minus 5, I will start laughing
58	Jo	these are/ the longer they are the/
59	Ea	you're saying, these are early early and two a/ five bases away is
61	Jo	naw, first, I let my gels run longer
62	Ea	haha
63	Jo	and besides with me these are 309, and this is 305 to 310 ((points to his film))
65	Ea	ah, but they are running the wrong way!
66	Jo	they run on the same level!
67	Ea	((ironic)) 305 and 309 run on the same level, right!
69	Jo	((impatiently)) listen, this up there is 520
70	Ea	uhuh
71	Jo	and hence this would be, if you take this to be 404, approximately 450. This means you would/
74	Ea	naw, might as well be 480. It starts there/
75	Jo	but you would thereby def/ thereby require, that you have 70/ a difference of 50 to 70 basepairs <u>because of</u> the salt!
78	Ea	man, this has something to do with the length of the gel run!
		((Etc.))

Note that what you see on an autoradiograph display depends on "what makes sense" in terms of experimental conditions and theoretical presuppositions. For example, in ln. 75 Jo objects against Ea's claim that the length of a band (its position on the film) is 480 by referring to the magnitude of the effect certain experimental variables (salt) should make. Note also that the point of such adversarial dialogue is not, as one might assume, the persuasion of one participant by another or the negotiation of firmly held opinions until a compromise is reached. First, participants develop their contributions as they go along in response to problem features they become aware of; they may not hold the respective opinions in advance. Second, the purpose of these exchanges appears not to be to reach an agreement among opponents, but to use their disagreement to produce novel (not previously obvious) features of the phenomenon discussed. For example, there is little effort on the part of participants in these exchanges to reconcile their differences. More generally speaking, there exists in these situations a *preference for disagreement* in contrast to the preference for agreement students of verbal encounters in other institutions, for example in doctor-patient interactions, have found.[17] Significantly, many adversarial exchanges do not end with an agreement but nonetheless produce a conclusion on which participants can proceed. Furthermore, even when an agreement is reached, this does not mean that the problem has been solved, as illustrated by the frequency of what one might call "negative solutions" – ways of undoing the problem without solving it. For practical purposes, results can be achieved which do not require a solution to the conceptual problem involved. Examples of such forms of remedial measures are proposals for different kinds of redressive action, such as for not showing the problem in a publication. Remedial measures are often proposed as free-standing solutions, that is as solutions which are not logically derived from the preceding exchange.

In sum, all patterns discussed above are inference producing devices that are interactionally accomplished, and they are initiated when the inquiry into the geography of the image collapses because bands are missing, occur in the wrong places, or display some other peculiarity which cannot be readily explained. Film talk begins as outlined in the projected structure. Indeed, the best indication of the relevance of the projected structure is participants' continued attempt to implement this structure (for example, they do not start by asking where the problem lies). But the final form of the exchange may look as in the following sample conversation:

OPENING SEQUENCE
Film is presented to a recipient
INQUIRY SEQUENCE
Q – A Inquiry about *kind of data* on film
Q – A Inquiry about *marker*; Problem: marker not seen
Oppositive exchange

(Argument about the appearance of the marker terminated by the next answer)

– A   Author offers second marker as an identifiable alternative

Q – A   Inquiry about length of 2nd marker; Problem: still unclear

– A   Author offers *probe* and length of probe as identifiable

O p t i c a l   i n d u c t i o n
(Attempt to derive length of marker-bands and probe-bands from visual inspection; 1st round)

Q – A   Inquiry about the *kind of constructs*; Problem: bands not visible

P r o c e d u r a l   i m p l i c a t u r e
(Reconstruction of procedures used in RNA preparation ending with performance recommendation)

Q – A   Inquiry about how the fragments were "cut"; Problem: recipient offers alternative answer, namely that bands are *starts*

O p p o s i t i v e   e x c h a n g e
(Arguments about whether certain bands are probe that is partially cut or secondary starts; no agreement)

INTERRUPTION

(Performance recommendation)

JOINING OF ANOTHER SPEAKER

Q – A   Inquiry about the length of marker/probe; Problem: length still unclear

O p t i c a l   i n d u c t i o n
(Attempt to determine length of marker-bands and probe-bands by visual inspection and reference manual; 2nd round; ending with performance recommendation re.probe)

INTERRUPTION

P r o c e d u r a l   i m p l i c a t u r e
(Reconstruction of length of marker-bands)

INTERRUPTION

(Performance recommendation regarding marker)

Much of what goes on in the exchange should be self explanatory from the above summary representation. The first "diversion" from the inquiry (*Q*uestion – *A*nswer) sequence comes about when the newsrecipient, who had been summoned to join in the inspection of the film by the author holding the film under his nose, rejects the latter's account of why the marker is not visible. The diversion is a short adversarial episode about whether the author's account is warranted. The inquiry sequence resumes when the author answers the marker-question indirectly by pointing out that there is a second marker on the film which offers no problems, but is arrested again by the question for the length of the marker, which the author cannot answer. He offers instead that he knows the location and length of the probe, thereby anticipating the next question in the sequence and initiating the second diversion: an attempt by both participants to infer the length of the marker-bands and the bands of the probe by going back and forth between these bands. A question about the kinds of construct inserted in certain lanes, which is prompted by participants noticing the absence of expected bands, briefly returns the dialogue to the inquiry sequence,[18] and then thrusts it back into another side sequence, when the recipient thinks the answer raises procedural problems. The next diversion from the inquiry sequence, a longer oppositive episode, follows suit: the recipient proposes an answer which differs from the author's, and the latter objects. Both this and the previous side sequence end with performance recommendations. In the final section, another member joining the exchange sets off round 2 of the length of the marker-and-the-probe induction performed on the film. When the author tells him he does not know the length, round 1 (2nd side sequence) turns out to have provided only a provisional answer. In the end, round 2 appears not to be definitive either; it terminates with performance recommendations for further inquiries necessary to satisfy this question.

To sum up this section, we want to draw attention to several features of the overall pattern of interaction in film analysis exchanges:

i. The inquiry sequence, heart of the projected pattern and glue for the actual one, remains incomplete. It may be completed in other rounds of talk, but single exchanges tend to get stuck in one of the side sequences;

ii. All inserted problem discussions (side sequences or diversions from the "projected path") tend to have recognizable closures, often performance recommendations;

iii. The "inserts" or "diversions" which split up the inquiry sequence constitute the bulk of the exchange;

iv. Problems emerge interactionally when questions cannot be readily answered or when answers are objected to by another party to the exchange;

v. Pauses and interruptions when other speakers raise a different topic have structural significance. They mark possible transition points to other patterns, precede conclusions and the like;

vi. Participants appear to accept specifications they have worked out only provisionally, as indicated by the fact that there is always the possibility that they may return to the same issue (say the length of the marker and the probe) at a later occasion and work through it in another round of shop talk.

6. Analyzability

What is at stake in these verbal exchanges is the analyzability of the visual image and not, as one might assume, the fit between previous theoretical hypotheses and the data obtained. What participants do when they talk is to negotiate the identity of the thick and thin bands or the blank spots on the film, by examining the features of the experiment which make the film analyzable. Analyzability is not just imposed upon the visual record by labelling the record and other techniques. Rather, it is *built into* the record from the beginning, through the way the experiment is designed.[19] In the case of autoradiographs of electrophoresis gels, these built in features include comparative standards such as:

i. Markers of length; i.e., fragments with known patterns which serve as a kind of measuring stick for the length of the resulting DNA and RNA fragments.

ii. A "blue marker"; which is a blue stain added to all lanes of the gel. When the gel is run, the blue signals, by appearing at a particular position, indicate that the separation of fragments in the reaction mixture has occurred.

iii. An "internal standard"; that is, a fragment inserted into all slots of the gel to assure the comparability of various parts of the experiment and to allow for the quantification of results.

iv. Finally, additional lanes may include familiar constructs whose patterns are known for purposes of comparison.

In addition, there will be positional clues such as slots or pockets which indicate which substance has been run through which column of the gel, and glued-on labels manually transferred from the electrophoresis plates to the film. All of these result in a kind of grid reminiscent of Dürer's drawing machine,[20] a reading grid designed to fix (make readable) the signal within the matrix it provides. (See Exhibit 3)

The grid formed by the markers, known constructs, internal standards and positional clues of an electrophoresis gel does not consist of a system of geometric coordinates like Dürer's machine but of in vivo biological specimen reactions: it is *of the same order and kind* as the traces obtained from the signal, and part of the *embodied optics of the experiment* conducted. As a consequence, the grid itself must first be read – the marks it creates on an autoradiograph film must be positioned and identified – and this as we found in the last section proves to be as problematic as the identification of the signal itself. In fact, there is no difference whatsoever in kind between the visual work necessary to identify the grid and the work required by the signal, and scientists make no distinction between these classes of variables as they work through the film. But why does the identification of the markers and comparative standards, which are included in the gel *to help locate and fix the signal* in the matrix they provide, create such a problem? Why, more generally speaking, is *analyzability* the problem and not, or in any case not at the bench, the theoretical meaning and interpretation of the data obtained?[22] Participants blame a variety of occurrences during the experiment for this situation. The most common are:

1. *Mix ups*: markers of length may get mixed up so that participants do not know which marker has been inserted into which

Exhibit 3. Reproduction of Dürer's picture of "The drawing grid" (*Source*: Jeaggli and Steck, 1969). In looking over a nodge and through the grid at his object, the painter can presumably locate every detail of what he sees on the grid of geometric coordinates. By transferring the resulting points to their equivalent location on the grid of the drawing table the painter creates an "objective" and "exact" rendering of the object

lane, or one of them may even have been forgotten. Lanes with different substances in the gel matrix may get mixed up, or the substances may not have been inserted in proper sequence.

2. *Manipulation problems*: The internal standards, the markers or known constructs included for comparison, may not have been "hot enough" (not radioactive enough), which renders them invisible or ambiguously visible compared with other bands. Known constructs may not have been "pure", causing them to suddenly yield patterns of bands different from those expected and documented. Bands may not appear or may spill into each other because the film has been exposed for too short or too long a time. When the film has been exposed for too short a time, the pockets on top of the gel which mark different lanes are not visible on the film, which makes it difficult to tell the top from the bottom of the film. The blue marker added to all lanes may run all the way through the gel and spill into the buffer solution when the apparatus is not turned off in time.
3. *Apparatus problems*: The voltage field generated during electrophoresis may not have worked properly, thus causing the bands on the film to deflect. As a result, it may no longer be clear which bands lie on the same level (have the same length). The plates between which the gel is inserted may break and part of the gel may become torn off.

Though some of these problems would in principle be avoidable, in practice they occur routinely. It appears that participants' practices are governed by principles other than those desirable from a methodological or epistemological point of view. For example, mix ups become understandable if one considers that, for reasons of "time" and in response to various demands of expediency, participants frequently handle 2–3 gels simultaneously, each of which displays approximately 20 different substances. Conceivably, some of the problems which occur could be eliminated by as simple a measure as the replication of the procedure, but, in practice, scientists attempt to use the results despite of the problems they exhibit. Why? They may not have the materials needed for a replication readily available, or may not have the time to obtain the materials and perform the work. And they cite the fact that any replication brings with it the danger of further problems. In practice, it seems, participants prefer a principle of variation over replication.[22] If the procedure has to be repeated, or so scientists argue, one might as well try out some variations which conceivably offer an improvement upon the previously obtaining situation.

Not all of the difficulties are "avoidables". To give a simple example, whether the marker is hot enough (radioactive enough and hence adequately visible on the film) depends on the strength of the other signals obtained. If they turn out to be weak then the marker will appear too hot. But the strength of the signals cannot be predicted precisely in advance. It is part of the experimental question to obtain information about the strength of the signals. Thus, to optimize the procedure, information would be needed that is contingent upon the outcome of the experiment, yet to obtain unequivocal results, the procedure would have to be optimized. Unambiguously visible data, whether signal or reference variables that make up the grid, are likely to be unattainable in this situation.

How is this ambiguity eliminated? The case analyzed in the last section suggests that the procedure is one of *embedding*. The details of the grid are identified by reference to the visibilities the grid variables display relative to each other, to their procedural history and to the experience of other participants. The shape and boundaries of the signal are identified by reference to the grid and equally to historical and experiential matters. The contexts invoked by these references constitute a web of meaning within which the data become fixed. In terms of Campbell's (1986) analogy of the cup and saucer it would appear that it is not the unambiguous entitivity of these objects which "edits" their linguistic designations. Instead, scientists proceed as if they were identifying cup and saucer by determining that the occasion is one of a tea party, and it is working this out which makes them fall back on talk.

7. Evidence

Is anything to be gained by moving from data to "evidence"? As indicated before, data become evidence, i.e., the data included in scientific texts, only after they have undergone an elaborate process of transformation. Significantly, the autoradiographs displayed in molecular genetics' papers are not identical to the troubled images (data) on which the work of seeing is performed in the laboratory. Further work is needed to arrive at figures that are self-explanatory and self-evident, as required by the research group's director; figures whose meaning is recognizable without consulting the accompanying text, figures which carry meaning on the face of them. What further work? Nearly all published images are carefully edited *montages* assembled from fragments of other images.[23] The original images are sometimes different exposures of an autoradiograph film taken from a single experiment, and at other times exposures from different experiments or runs of a gel. The resulting montages display at least three analytic orderings:

1. They rely on the *methodical production of a perspectival* (3-dimensional) *order* which puts the signal into the foreground and the "noise" into the background. The activities through which this perspectival order is construed are mundane. They consist of the following practices:
   – cutting off, at the top and bottom of a visual display, bands considered as artifacts, as unclear, or simply as irrelevant to the "message" to be conveyed;
   – manipulating the exposure time of films or photographs of films to enhance the visibility of bands judged to be significant, and to decrease the visibility of unexplained traces;
   – selection of the lanes from several runs of gels which best display the features proposed. These lanes are cut out (some scientists insert an additional marker in the middle of the gel in order to have a lane in which to cut) and glued together.

The following exhibit shows a montaged autoradiograph upon which the above manipulations have been performed. The lines inserted mark the cuts and the clippings from which the present figure was assembled. (See Exhibit 4)

The result of the above manipulations is a montage of relatively "clean", "pure" or "beautiful" signals according to aesthetic criteria which specify, in an area of research, what counts as a

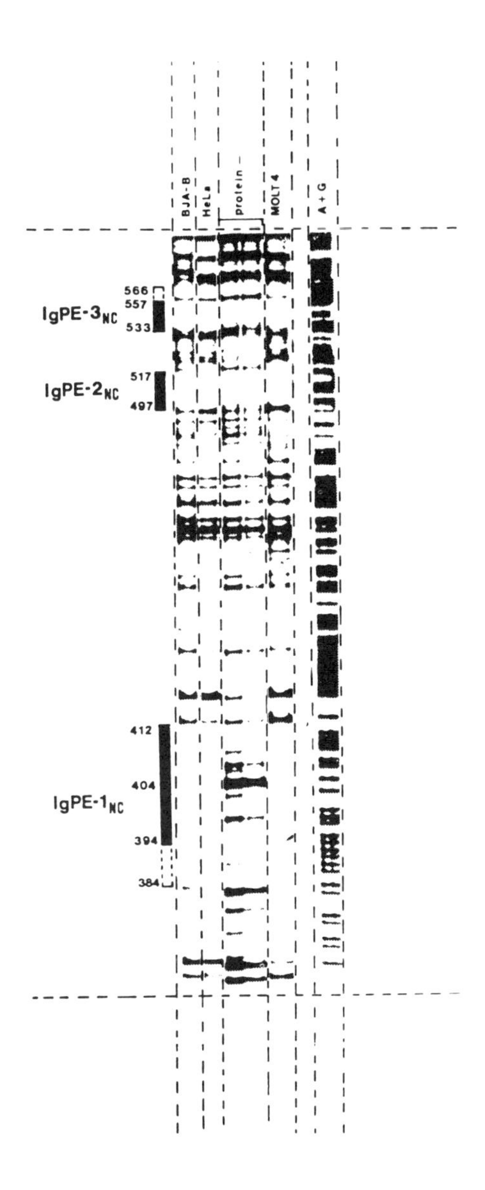

Exhibit 4. Montaged autoradiograph display assembled from various original films. The lines indicate the cuts participants made in piecing the display together. Note also the pointers (bars) and labels at the edges of the image. Compare this autoradiograph with the unedited film reproduced in Exhibit 1

"good" figure. Note that such judgments are not extrinsic to scientists' judgments on *what* the figure reveals.

2. The second ordering imposed upon the film displays the signal *within a matrix of other signals*. This manner of exhibiting the signal provides for the analyzability of results in terms of comparisons between signal and reference lanes, and thereby for an evaluation of the consistency and coherence of results. This ordering is achieved by some of the same methods as the perspectival order: fragments of bands from different runs or experiments are cut out and glued together *as if* they were emanating from the same experiment and run, the embodied optics of the gel run is *reconstituted* through the careful composition of traces in a documentary display. Some of the vertical cuts in Exhibit 4 (marked by lines) indicate such fragments. Thus, the display format is that of the grid described earlier, with the difference that the grid is the post hoc result of image composition.
3. Finally, autoradiograph figures composed to present "evidence" rely on the use of "*pointers*". These are marks added to the image which suggest a particular reading of the display by indicating some features as significant and ignoring others. Typically, they consist of arrows, brackets, lines or other visual clues inserted at the boundaries of the image. Additional aids in inducing desirable readings of a figure are of course the title and written explanations which frame the image.

Now to avoid misunderstanding, let us stress that we are not suggesting here that the evidence thus created is purely fictional – however fabricated it may be. But neither does it correspond to the "data" or signals obtained in the laboratory. Rather, this montage is a members' way of visually reproducing *the sense of "what was seen"* which is an *upshot* of participants' shop talk negotiations; an accomplishment of – not a precondition for – their work. Talk attached to (in the sense of Section 4) visual materials was crucial insofar as it provided participants with candidate formulations of the reality they "saw".

A final note. It is important to realize that the aesthetically enhanced, montaged version of this reality – evidence – is not the end of the story. Evidence, i.e., visual objects that have been trimmed and fixed, are retransformed into "data" when these objects are critically inspected by an audience, the readers of the evidence. In other words, evidence tends to become problematic once more when it is seriously considered by informed scientists or competitors in the field. The inspections performed by participants on published visual displays provide an occasion in which the flexibility of visual objects is highlighted once again. Images (visual evidence) do not function in the literature in the way one might assume; that is, by reducing the indexicalities of the text, by displaying the data unequivocally, by adding the certainty of proof which the text can only refer to, but not "show". Quite to the contrary. Images, perhaps more than texts, provide infinite opportunities for visual exegesis, thereby functioning to keep the discussion open, not closed.

8. Conclusion

To conclude this paper, consider the following summary of the process of evidence fixation. (See Diagram 2)

This paper has emphasized the difference between "data" and "evidence" as a means of distinguishing between different modes of practice through which visual objects are constructed in the stream of laboratory shop work. We have found that in the science studied, processes of seeing appear to be interactionally dissolved in shop talk, and we have focussed upon the conversational routines and inference machineries in terms of which seeing becomes socially organized in talk. Conversational inference devices are employed as participants run into problems in recognizing visual objects, in determining, that is, the identity of the black and white bands exhibited on autoradiograph films. With the help of these conversational devices, participants develop a sense of "what was seen" on these data displays. Through montage, this sense of what was seen is transformed into evidence. Both processes constitute what we have called the fixation of evidence. Evidence is the aesthetically enhanced, carefully composed rendering of flexible visual objects that, through the meandering interrogatory processes

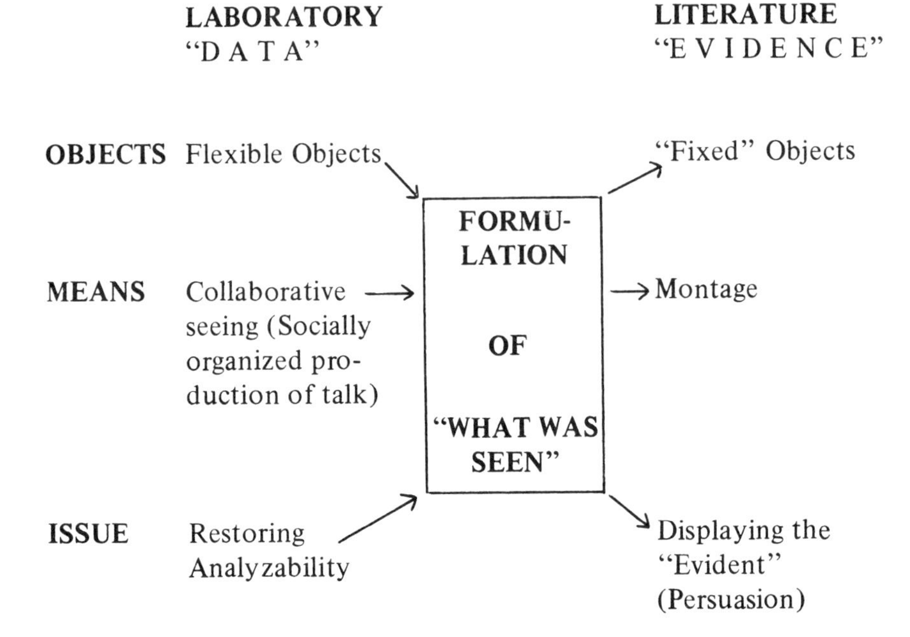

Diagram 2

of image analyzing talk, have been "embedded" and entrenched in procedural reconstructions, local experiences and in the landscape of the data display.

We have tried in this study not to miss the interactional work of seeing, through which proto-data – ambiguously visible unidentified data traces – are transformed into "data", and we have touched upon the practices through which these data are mounted as evidence in publishable papers. It remains for future research to demonstrate how published evidence is deconstructed and retransformed into questionable data when these papers are read; when the evidence is inspected by the wider audience in a specialty field.

Notes

1. See Grunbaum (1960) for a summary and critical discussion made by Duhem and Quine.
2. Laboratory studies are based upon *in situ* observations of scientists at work in their laboratory and upon other ethnographic methods. For some of the major laboratory studies published so far see Knorr Cetina (1981); Latour and Woolgar (1979); Lynch (1985a); Traweek (1988) and Zenzen and Restivo (1982). For a discussion of the assumptions and results of these studies see Knorr Cetina (1983). For an overview over recent developments in science studies in general see Knorr Cetina and Mulkay (1983).
3. These distinctions are embodied in scientific practice, but they are not made by practitioners in these terms. For example, scientists may refer to the proto-data inspected in the lab as "evidence", and they call the displays published in the literature "data". Moreover, the distinctions are fuzzy, not sharp. Our purpose in drawing a line between "data" and "evidence" is to draw attention to different modes of practices participants employ in regard to visual materials. It is not to propose a taxonomy of these materials.
4. For an introduction to electrophoresis methods and to autoradiograph data see any textbook in molecular genetics, for example Alberts et al. (1983).
5. This is not to say that "enhancements" may not become the subject of sustained inquiry if practitioners feel that such attention is warranted. However, such attention is the exception rather than the rule. When it occurs, it can prove to be important, and turn around a whole line of inquiry.
6. For an earlier laboratory study that deals with scientists' shop talk on a more general level see Lynch (1985a). For a discussion of the characteristics of shop talk as a communicative form in the sciences we currently study see Amann and Knorr Cetina (1988b).
7. For a detailed description of processes of decision making and consensus formation in the laboratory see Amann and Knorr Cetina (1987).
8. For an attempted rebuttal of this finding see Eglin and Wideman (1986).
9. By conversation analysis standards (e.g., Sacks, Schegloff and Jefferson 1974:731–34) the following data are fairly grossly transcribed. We have neglected overlaps and omitted indications of the length of pauses, and we have not transcribed explosive aspiration, "latching" or prolonged prior syllables. We believe, however, that the transcriptions are adequate for the level of analysis we attempt in this paper, and we hope that they are easier to read for the audience to whom the paper is addressed. The following transcribing conventions were used:

   / "Interruption".
   ( ) Single parentheses indicate the transcriber was not sure about

the words contained within parentheses. Empty parentheses indicate talk inserted in or before passages relevant to the case presented.

(( )) Double parentheses indicate comments by the transcriber.

? Rising intonation.

10. The notion "side sequence" does not adequately capture the fact that "diversions" from the inquiry sequence dominate in real time film analysis exchanges.
11. As judged from the fact that the patterns of interactional organization found may not be the most efficient way of solving the problems posed, and by the fact that other patterns (other passages to the goal) can be readily imagined.
12. For an interesting study of how the production of inferences relies on vehicles other than "thought" in everyday situations see Lave (1987).
13. The name of the pattern borrows from Grice's notion of "conversational implicature" (1975) mentioned before and from Cicourel's work on "procedural knowledge" (e.g., 1974; 1975; 1978).
14. The expression borrows from Giddens (1984).
15. We are indebted to Karl Heinrich Schmidt for the name of the pattern.
16. For an example of this form of optical induction, see Amann and Knorr Cetina (1988b).
17. For a different finding in regard to science see Lynch (1985: Chs. 3–5). The notion "preference for agreement" is used in two ways in the relevant literature: On the one hand, it refers to formal agreement, as in sentences beginning with "Yes, but..." which are usually polite versions of disagreements. On the other hand the notion also refers to a more general tendency to express agreement with a speaker's utterance and to keep to oneself possible disagreements with his/her opinion. We are using the notion in the *latter* sense. See Brown and Levinson (1978), Pomerantz (1975) and Sacks (1973).
18. In this case the problem arose out of the visual work performed in the side sequence rather than out of a question posed in the inquiry sequence. This and other idiosyncracies of image analyzing exchanges are not atypical. Rather, they show that real time film talk is often more messy than suggested in Section 4 of this paper. More messy in the sense that it includes such idiosyncracies, but also "orderly" in the sense that the idiosyncracies exhibited can usually be explained. In the present case, visual work on the film naturally raises the possibility that things will be noticed (and subsequently pursued) which have nothing to do with the original goal of the optical induction.
19. For a similar finding in regard to a neurosciences research project see Lynch (1985b:52). Lynch talks about pre-linguistic modes of order production built into visual records, and about "endogenous" geometries of cellular material being exposed and brought into alignment with "exogenous" graphic formats. We are concerned with the "pre-docu-

mentary" optics of the experiment that embodies scientists' concern with the analyzability of their data.

20. Dürer's grid had a slightly different purpose, as the figure-legend indicates. For a detailed interpretation of Dürer's grid, see among others Kutschmann (1986).
21. Theoretical interpretations may become a problem when participants write up their results, when they discuss what to say and what data to include in the papers they produce.
22. In this paper, we cannot discuss the issues raised by these observations. For more details see Amann and Knorr Cetina (1987). For a discussion of replication in experiments in physics see Collins (1975).
23. For an interesting paper on how natural objects are made visible through montages see Lynch (1985b).

References

Albert, B., Bray, D., Lewis, J., Raff, M., Roberts, K., and Watson, J.D. (1983). *Molecular biology of the cell.* New York and London: Garland.

Amann, K., and Knorr Cetina, K. (1987). *Consensus formation in the laboratory*. University of Bielefeld, Bielefeld, Mimeo.

Amann, K., and Knorr Cetina, K. (1988a). Thinking through talk: An ethnographic study of a molecular biology laboratory. In R.A. Jones, L. Hargens and A. Pickering (Eds.), *Knowledge and society: Studies in the sociology of science past and present*, Vol. 8. Greenwich, CT: JAI Press.

Amann, K., and Knorr Cetina, K. (1988b). Werkstattsgespräche in der Wissenschaft: Am Beispiel der Molekularbiologie. In H.G. Soeffner (Ed.), *Sprache und Gesellschaft*. Frankfurt/M.: Campus Verlag.

Brown, P., and Levinson, S. (1978). Universals in language usage: Politness phenomena. In E. Goody (Ed.), *Questions and politness*. Cambridge, MA: Cambridge University Press.

Cicourel, A. (1974). Interviewing and memory. In D. Cherry (Ed.), *Pragmatic aspects of human communication*. Dordrecht: Reidel.

Cicourel, A. (1975). Discourse and text: Cognitive and linguistic processes in the study of social structure. *Versus: Quaderni di Studi Semiotica* (September – December).

Cicourel, A. (1978). Language and society: Cognitive, cultural and linguistic aspects of language use. *Sozialwissenschaftliche Annalen* 2:325–359.

Collins, H. (1975). The seven sexes: A study in the sociology of a phenomenon, or the replication of experiments in physics. *Sociology* 9:206–224.

Campbell, D. (1986). *Models of language learning and their implications for social constructionist analyses of scientific belief.* Paper presented at the Annual Meeting of the Society for Social Studies of Science, Pittsburgh, Pennsylvania.

Eglin, P., and Wideman, D. (1986). Inequality in professional service encounters: Verbal strategies of control versus task performance in calls to the police. *Zeitschrift für Soziologie* 15(5):341–362.

Giddens, A. (1984). *The constitution of society.* Berkeley: University of California Press.

Goffman, E. (1971). *Interaction ritual.* New York: Anchor Books.

Gombrich, E.H. (1960). *Art and illusion.* Princeton, NJ: Princeton University Press.

Grice, P. (1975). Logic and conversation. In P. Cole and J. Morgan (Eds.), *Syntax and Semantics*, 3: *Speech Acts.* New York: Academic Press.

Grunbaum, A. (1960). The Duhemian Argument. *Philosophy of Science* 27(1):75–87.

Jaeggli, A., and Steck, M., Eds. (1969). *A. Dürer. Die kunsttheoretischen Werke* Bd. 1: *Underweysung der Messung, Nürnberg 1525.* Zürich.

Knorr Cetina, K. (1981). *The manufacture of knowledge: An essay on the constructivist and contextual nature of science.* Oxford: Pergamon Press.

Knorr Cetina, K. (1983). The ethnographic study of scientific work: Towards a constructivist interpretation of science. In K. Knorr Cetina and M. Mulkay (Eds.), *Science observed: Perspectives on the social study of science.* London and Beverly Hills: Sage.

Knorr Cetina, K., and Mulkay, M., Eds. (1983). *Science observed: Perspectives on the social study of science.* London and Beverly Hills: Sage.

Kuhn, T. (1970). *The structure of scientific revolutions.* Chicago: University of Chicago Press.

Kutschmann, W. (1986). *Der Naturwissenschaftler und sein Körper.* Frankfurt/M.: Suhrkamp Verlag.

Latour, B., and Woolgar, S. (1979). *Laboratory life: The social construction of scientific facts.* London and Bevery Hills: Sage.

Lave, J. (1987). *Arithmetic practice and cognitive theory: An ethnographic inquiry.* Berkeley: University of California Press.

Lynch, M. (1985a). *Art and artefact in laboratory science: A study of shop work and shop talk in a research laboratory.* London: Routledge and Kegan Paul.

Lynch, M. (1985b). Discipline and the material form of images: An analysis of scientific visibility. *Social Studies of Science* 15:37–66.

Merleau-Ponty, M. (1962). *The phenomenology of perception.* London: Routledge and Kegan Paul.

Mulkay, M. (1979). *Science and the sociology of knowledge.* London: George Allen and Unwin.

Pomerantz, A. (1975). *Second assessments: A study of some features of agreements/disagreements.* Unpublished Ph.D. Dissertation. Division of Social Sciences, University of California, Irvine.

Quine, W.v.O. (1960). *Word and object.* Cambridge, MA: MIT Press.

Sacks, H. (1973). *Current research in conversation analysis: The Preference for agreement.* Paper presented at the Summer Institute for Linguistics, Ann Arbor, Michigan.

Sacks, H., Schegloff, E.A., and Jefferson, G. (1974). A simplest systematics for the organization of turn-taking for conversation. *Language* 50:696–735.

Traweek, S. (1988). *Buying time and talking space: The culture of the particle physics community*. Boston: Harvard University Press.

West, C. (1983). 'Ask me no questions...'. An Analysis of queries and replies in physician-patient dialogues. In S. Fischer and A. Todd (Eds.), *The social organization of doctor-patient communication*. Washington: Center for Applied Linguistics.

Zenzen, M., and Restivo, S. (1982). The mysterious morphology of immiscible liquids: A study of scientific practice. *Social Science Information* 21:447–473.

Time and documents in researcher interaction: Some ways of making out what is happening in experimental science *

STEVE WOOLGAR

Department of Human Sciences, Brunel University, Uxbridge, Middlesex UB8 3PH, UK

1. Introduction

This paper is based on arguments recently developed for the strategic importance of treating scientific work as "practical reasoning". Although it is not appropriate to repeat these arguments in full, it is worth recalling that "practical reasoning" is intended as a generic term for a variety of social processes whereby practitioners effect connections between what are taken as "surface documents" (which might take the form of signs, marks, indicators, utterances, actions, gestures and so on) and the "underlying reality" (which might include, for example, "what the mark shows", "what motivated that action", "what gave rise to this utterance", "the circumstances which render that gesture sensible" and so on). In its general formulation, the notion of "practical reasoning" owes much to early work in both ethnomethodology (particularly Garfinkel's, 1967, discussion of the documentary method) and structuralism (particularly discussions of the signifier-signified relationship as found, for example, in Saussure,

* An earlier version of part of this paper was read at a conference on 'Communication in Scientific Research', Simon Fraser University, Burnaby, B.C., Canada, 1–2 September 1981. Valuable transcription assistance by Doug McLaughlan is gratefully acknowledged as are helpful comments by Mark Boardman. The research of this paper was supported in part by a grant from McGill University Faculty of Graduate Studies and Research. Thanks are also due to the members of the University laboratory who generously accommodated the study which provided material for this paper.

1916; and Sturrock, 1979). Here, I investigate how scientists accomplish connections between specific research documents and their underlying realities. I will argue that such connections are assessed as "good enough" or "practically adequate" in the course of scientific work, despite philosophical and methodological arguments that such connections are in principle inadequate, indefensible, and even impossible. The possibility of such arguments constitutes a methodological dread, a Pandora's box of horrors, which scientists skirt around, avoid or ignore in the course of their practical actions. The study of practical reasoning in science is here construed as the study of scientists' artful management of possible methodological horrors.

This formulation of our analytic perspective will bear considerable refinement in the light of the empirical analysis below.[1] For now, let me merely re-emphasise that my reference to scientists' *artful* practices is not intended to imply any kind of deceit on the part of our long suffering subjects, but merely to underscore the craft-based skills of laboratory work.

This investigation draws upon materials from eighteen months participant observation of the work of a group of solid state physicists. My general concern is to document the sense in which the artful management of methodological horrors is an interactional accomplishment. In particular, I wish to consider how various management practices yield determinations of 'what is happening' in the course of an experiment.

I begin with a brief 'background description' of the work of the laboratory. This can serve both as an introduction to the instance of experimental practice subsequently analysed in detail, and as an occasion for raising some general issues about the adequacy of description. My own attempt to provide a version of 'what is happening in the laboratory' is presumed no less liable to the methodological horrors than the scientists' efforts to determine 'what is happening in the experiment'. Without assuming their congruence, we can draw on issues and problems arising in our own work as a heuristic for examining the management of similar issues and problems in the work of the laboratory.

2. An initial description of the work of the laboratory

One main interest of the group of solid state physicists I have been studying is the behaviour of amorphous metal alloys (so-called metallic glasses). In their amorphous state (corresponding to irregular atomic alignment), metal alloys have a number of unusual properties with potential for industrial application. For this reason, much research in amorphous metal alloys has been carried out by industrial scientists with a view to developing, for example, more efficient high voltage transformers.

Amorphous metal alloys are, however, inherently unstable in that heating to a sufficiently high temperature or even isothermal annealing (that is, maintaining the alloy at the same temperature for a sufficient time) brings about an irreversible transformation into the crystalline state (a state of more regular atomic alignment where the atoms assume a dense packing configuration – Altounian et al., 1981). The change to a crystalline state is accompanied by marked changes in the physical properties of the metal alloys. In order to take advantage of the properties of the amorphous metal alloy, it is important to understand the conditions and mechanisms which effect the crystalline transformation. Of particular concern is that crystalline transformation of some alloys seems to occur at or near room temperature. I was told, for example, that several samples of CuZr (copper zirconium) had crystallised after being subject to the annealing conditions of the desk drawer in which they were kept!

Samples of amorphous metal alloy are prepared in the form of strips of shiny ribbon (typically 1/4 inch wide) by a technique known as 'spin melting' (Chaudhari et al., 1980). This technique was adopted in the laboratory after it was decided that the samples supplied from elsewhere were of unreliable quality and purity. Small amounts of the constituent metals are initially fused together by arc melting in argon. Subsequently, fragments of the resulting alloy (typically 1.5 grms) are again melted and pressure-injected onto a rapidly spinning copper wheel. The molten alloy cools rapidly as it makes contact with the copper wheel, thus forming ribbons of amorphous alloy.

Short lengths (typically 2 inches) of the (presumedly) amorphous metal alloy ribbon are then subject to various kinds of experiment, the general objective of which is to determine the nature of the structural changes which amorphous alloys undergo in the course of heating (or cooling). These changes are monitored by reference to three main properties of the alloys: electrical resistance; magnetic susceptibility; and X-ray diffraction. Although interpretations of structural changes frequently draw on one or more indicators, each indicator requires a separate experimental technique. Here we shall concentrate on the monitoring of electrical resistance. Under this rubric two main kinds of experiment are carried out: isothermal annealing and heating at a constant rate. In the latter case, a small sample of the ribbon is placed in a thermostatically controlled oven and subjected to heat. As its temperature changes, changes in its electrical resistance are monitored in the form of a trace on a pen-chart recorder. It is this last case of laboratory work I want to focus on in detail.

3. The historicity of experiment

The materials I wish to consider comprise my observations (in the form of field notes), various graphs and charts, and a detailed transcription of talk during one of the experimental runs referred to in my initial description of the laboratory. This run, an investigation of changes in the electrical resistance of a sample of copper zirconium (Cu50Zr50) as it is subject to heat, took place during the morning of Thursday, 22 January 1981. As we shall see, it is important for understanding the interpretive work involved to note that this was the fifth in a series of runs using samples of copper zirconium of different composition (Cu x Zr 100–x, where x varies between 0 and 100). The talk was tape-recorded using a hand-held Sony TCM-260 cassette recorder with an inbuilt condenser microphone.[2] The entire stretch of recorded talk lasts a total of 72 minutes 31 seconds, and the transcription runs to 28 pages and approximately 1,600 lines. (Necessarily, then, the full transcript is not reproduced here.) This period covers the time between the initial setting up of the experimental run, when the oven containing the ribbon sample is flushed with helium gas (in order to rid the immediate surrounds of the ribbon sample of any contaminating oxygen), until just before the heat source is switched off and one of the main participants decides it is time to leave for lunch. The major interest for present purposes is that the talk takes place over and around the pen chart recorder; as the pen traces changes in the resistance of the alloy ribbon, it plots a line graph of resistance versus temperature. It would be difficult to claim that the talk is solely and continually concerned with the line graph. An inspection of the entire transcript reveals mention of topics as diverse as the weather, skiing, the ability and performance of undergraduates, British public schools and various television programmes. Nonetheless, the talk does appear to return again and again to the trace. I therefore concentrate on those aspects of the talk concerned with what the trace is showing, what it means, what it indicates is happening to the sample, how this compares with other traces and so on. In other words, the transcript of the talk will be treated as an occasion for construing the work done by participants in interpreting the document as it unfolds before them.

It is important to stress the temporal dimension of the interpretive work being done. The issue for participants is not just the meaning of a fixed trace; it is not simply the kind of retrospective assessment they engage in when looking at past results, at published graphs and so on. In this case, the talk is about a document *as it develops*. Thus, at any particular moment, a participant's interpretation of the trace may involve an assessment both of the recent past of the trace and of its possible future course. Indeed, prediction and anticipation are central to certain segments of this interaction. In order to capture this ongoing quality of the chart's interpretation, I kept a record of the point reached by the trace when selected utterances occurred. Since the chart recorder has a clearly graduated scale along its x-axis (1/10 inch divisions), it was possible to note the position of the pen when particular utterances were made. Figure 2 is a reproduction of the entire completed trace as it looked at the end of the run. Figure 1 represents the portion of the trace corresponding to the recorded talk; the points when certain selected utterances occurred is indicated on this figure by the equivalent line number of the transcription.[3] For example, line 1856 of the transcription is T's comment: 'It's *still* going see'; in Figure 1, 1856 indicates the point reached by the trace at the time this comment was made.[4] This system of

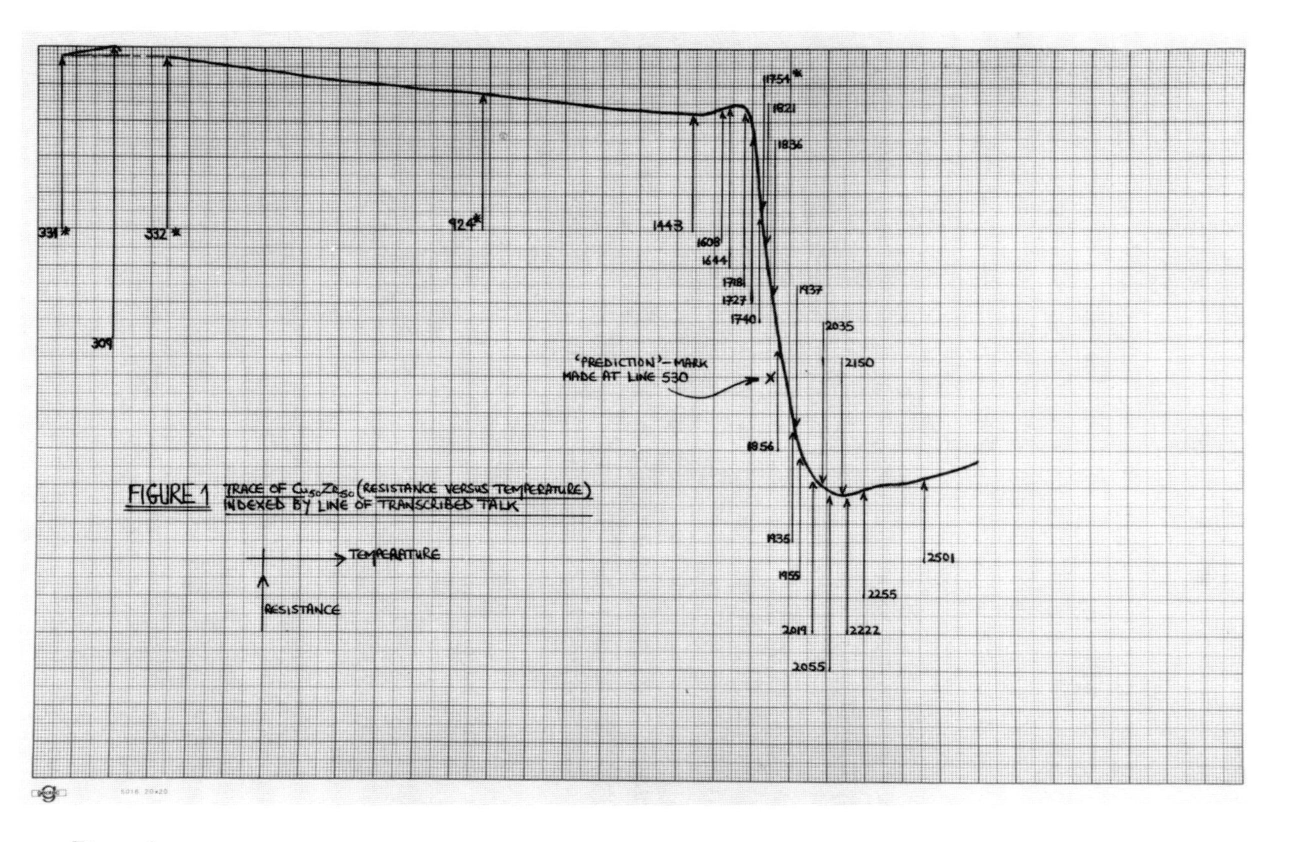
FIGURE 1 TRACE OF $Cu_{50}Zr_{50}$ (RESISTANCE VERSUS TEMPERATURE) INDEXED BY LINE OF TRANSCRIBED TALK
'PREDICTION'—MARK MADE AT LINE 530
TEMPERATURE
RESISTANCE
331*
309
332*
924*
1443
1608
1644
1718
1727
1740
1754*
1821
1836
1856
1937
2035
2150
1935
1955
2019
2055
2222
2255
2501

Figure 1

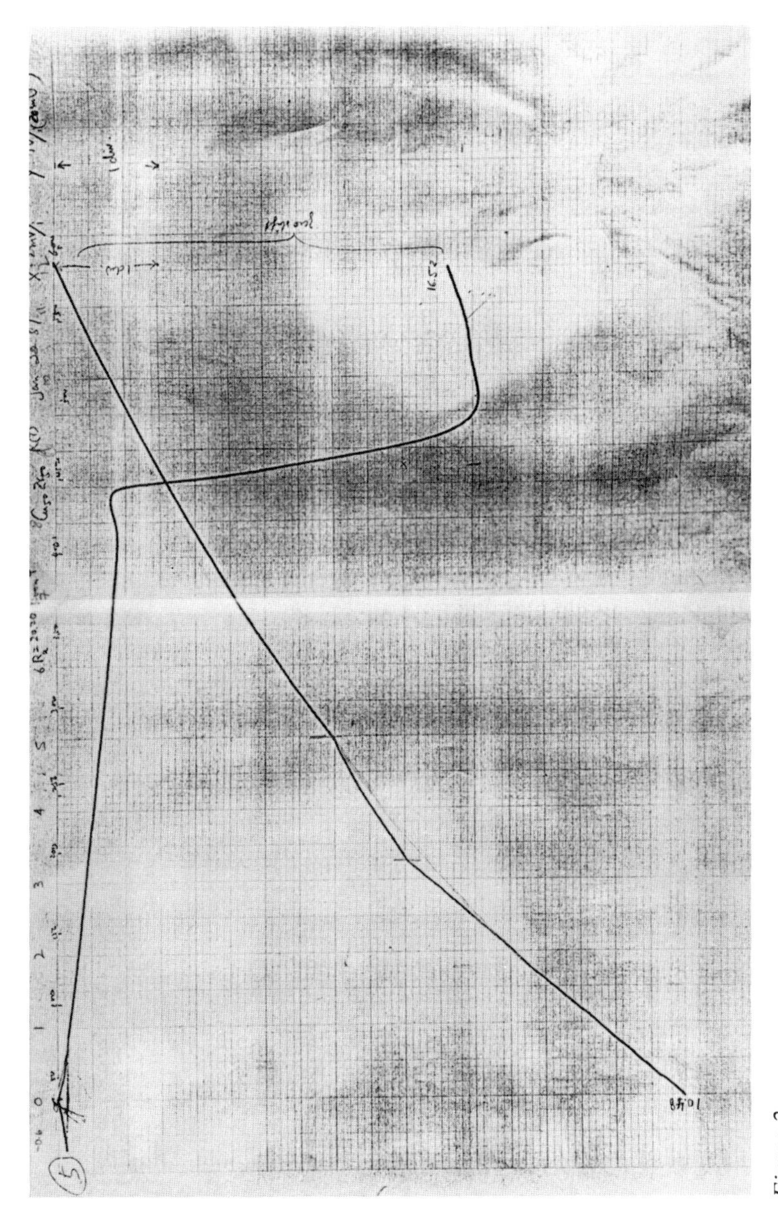

Figure 2

indexing makes it possible to correlate the incidence of talk with the development of the trace at the point at which different sections of commentary were advanced.

The only other attempt to analyse detailed contemporaneous materials of an instance of scientific interpretation is the discussion of the recorded talk between researchers in the course of their observation of the first optical pulsar (Garfinkel et al., 1981). The major drawback of that analysis is that the visual display confronting the researchers was not available for subsequent analysis; although the researchers were plainly responding to, and determining the significance of the display before them, there is no record of the (unfolding) display, and hence no way of investigating the ties between the researchers' interpretive (discovery) talk and the immediate visual foil for their interpretation. By contrast, the present materials provide an important opportunity for closely monitoring the ways in which participants' minute-to-minute assessment and interpretation of the trace is tied to their assessment and construal both of its relevant past and projected future. We can examine how participants constitute antecedent circumstances as having given rise to the current state of the trace. It is also possible to discern the ways in which the scientists select and attach differential significance to past events and their serial location such that the present direction of the trace is made sensible. Obviously, whether or not the preparatory procedure illustrated earlier has been successful (for example, whether and to what extent the resulting ribbon "actually is" amorphous) is the subject of complex interpretive work by participants. And these initial procedures are themselves a resource for subsequent argument about the nature of the sample and what is happening to it ("what it is doing"). Thus, the set of past events which can be invoked include things that might have happened in the production process. "The production process" provides a stockpile of scenes recapturable for present interpretive purposes. For example, a strangely low measure of electrical resistance may be explicable in terms of the alloy fragment having been cooled insufficiently quickly, thereby allowing the alloy partially to crystallise. Similarly, aspects of the procedure for finally preparing the alloy ribbon for experiment ("tagging up") were frequently cited as the origin of troubles encountered in the experiment itself. A trace was sometimes abandoned in mid-run because its character was said to reveal that the electrical contacts had been inadequately fixed on the ribbon. Indeed, some of the early exchanges in the present transcript (lines 101–133; 343–346) comprise participants' evaluation of the course of the trace vis-à-vis the immediately preceding run, which was abandoned when the trace was held to exhibit the symptoms of poor electrical contacting.

One general strategy for conjuring up the past is the enunciation of 'captions' so as to make visible and relevant a particular aspect of the preparatory procedure; by offering these 'captions', participants "tell" a relevant feature of the past. A remembered scene thus speaks to a previously unnoticed feature of what was there all along (had they but realised it). The strategy is analogous to my saying of a photograph (my "pointing out") that it is in fact a picture of x. The caption "it is in fact a picture of x" provides a way of "telling" the relevant features of the photograph.

As just mentioned, the invocation of relevant past sometimes resulted in the abandonment of a particular run. More consequentially, the combined invocation of several features of the past occasionally resulted in a major modification of laboratory technique. For example, the spin melting technique was adopted in the laboratory after it was decided that samples supplied from elsewhere were of unreliable quality and purity. In other words, the adoption of this technique was the result of the determination that a series of traces exhibited the presence of foreign matter in samples. Although the spin melting technique had been firmly routinised by the start of my time in the laboratory, it is possible to speculate how 'foreign matter' had been "detected". Each of a number of traces were taken to exhibit the same relevant feature, and this feature was held to be the identifying characteristic of a family of traces, which together were taken to establish that a faulty characteristic of the metal alloy originated in the production process of some industrial laboratory. In these instances, the upshot of finding a contingent past in the trace was dramatically consequential. In the materials below we focus on similar determinations of the past's relevance which had less dramatic results.

In sum, I suggest these material provide a useful basis for understanding the work done by participants in constructing what can be thought of as "histories-for-the-moment". More than this, they make it possible to pursue an interesting sense in which the interpretive work of the experiment is "historically bounded". Rather than attempting to understand participants' actions by using an analyst's version of "the prevailing historical context (circumstances)". I suggest that interpretation in the laboratory is deeply bound up with participants' own strategies for temporal assignation and construal of events, in short with their localised assembly of histories. I begin by elucidating some of the ways in which participants to the interaction make out the basic character of the trace as it appears before them. But first we need to introduce the participants.

4. Parties to the interaction

In the present interaction the most immediately recognisable participants are T and Z (post-doctoral fellows), B and J (faculty members) and myself (S). Apart from three brief absences by T (410–417; 1529–1623, 2521–2518) and one by Z (1855–1911), the two post-doctoral fellows stay near the experimental apparatus throughout the entire interaction. J comes into the laboratory on three separate occasions (236–237; 647–861; 1348–1630) while B appears only once (1833–1855). Even though the others, at no time, leave him alone with the equipment, S is the only person to remain by the apparatus throughout.

Further inspection of the transcript reveals that other participants also feature in the talk: Mario (529 – a graduate student), Frank 1413 – a technician), individual undergraduates (1535, 1553), specific colleagues (for example, 1711) and generalised other colleagues (for example, "those people" – 1826). None of these persons themselves contribute utterances to the transcript. Yet the invocation of each of these persons is itself a document which occasions interpretation. For example, individuals are commonly named in the talk as a way of introducing results, opinions, findings or other actions and the meaning and relevance of these things is subject to interpretation. When Z says (1711) "Giessen says it has something to do with oxygen," the statement – "it has something to do with oxygen" – is introduced as a tied knowledge claim. The inclusion of the modality – "Giessen says" invites assessment of the statement in relation to what might be known about its (alleged) proponent. The (transcribed) utterances of the immediate participants stand as documents in the sense that they are examinable for what was intended or meant by the utterer. At the same time, certain utterances appear to work as reports upon the actions, opinions, produced results and so on of other non-present persons; participants can conjure up the activities of others, the sense and significance of which is also available for consideration. And, of course, the matter does not stop there. For scattered around the laboratory, on benches, in nearby offices, on desks and in files are documents in the more usual (literal) sense: various output traces, journal articles, charts, jottings, notes, and so on, take their place along side the utterances, gestures, statements and claims of human participation in a general melee of documents. At any moment, various of these documents are selected, subjected to interpretation or invoked as argumentative resources. Of course, I assume the centre piece of the interaction is the trace appearing on the chart recorder: the "underlying reality" of this text is the main concern of participants' interpretive work. But if this particular document is the star attraction of the interaction sequence, from an interpretivist standpoint we are dealing with a cast of thousands.

The distinction between human and inanimate parties to the interaction should be treated with caution. Documents of various kinds come into play, but their connection with human agency varies according to the particular instance of interaction. For example, the sense of an experimental result (for example, an output trace) can be altered by redescribing its origins (for example, the differential scanning calorimeter) in terms of human involvement in those origins (for example, *Giessen's use of* the differential scanning calorimeter). Consequently, it is important to remember that the character of a document is an occasioned accomplishment. Certain parties to the interaction can claim to be able to speak on behalf of documents and this can have significant implications for the outcome of interpretive activity. The persuasiveness of a particular interpretation of a document might hinge on a successful ascription of origin. For example, to make out a document as the product of human agency allows for the possibility that motives, intentions and so on are involved in the production process. The main point is that the present episode is to be treated as an interaction between produced documents, rather than just between scientists. The audible presence of scientists' speech, its tape-recordable quality, should not lead us to overlook the sense in which the line on the graph also speaks.

5. The interactional adequacy of description

An immediately noticeable aspect of the interaction is that participants repeatedly "tell" features of the experimental run without concern for the principled problems of adequately establishing a connection between their utterances and its referent. A particularly vivid example is participants' assessment of the current temperature of the sample:

(A)
1642 Z whats the temperature now lets see
1643 (2.5)
1644 Z eighty two or something like that
1645 (3.5)

(Numbers in parentheses represent silences or pauses in seconds or tenths of seconds.)

"Eighty two" makes sense to me as Z's way of saying that the pen has moved 8.2 inches from its origin. This distance, expressed in tenths of an inch, passes as an adequate response to an inquiry about temperature. Z does not on this occasion find it necessary to specify temperature in degrees. By contrast, a few seconds later we find:

(B)
1745 T now shall we read er read the temperature?
1746 Z mm ok if you want to check
1747 T (we have to look at)
1748 T the left one?
1749 Z yeah yeah yeah read something anyway
1750 T seventeen thirty

1751 (2.0)
1752 T is it seventeen thirty? – seventeen twenty
1753 (1.5)
1754 T its seventeen twenty thats – one two three four five six seven
1755 eight sixteen seventeen twenty – okay a little bit difference
1756 Z (not bad)

(Note that lines 1747 and 1756 occur simultaneously with utterances in the prior lines. Parentheses indicate that the utterances are unclearly heard by the transcriber.)

At line 1750, T reads the left of two digital displays on the experimental console and then compares this reading with the traverse of the chart pen across the x-axis of the graph (lines 1752 and 1754–5). In this instance, temperature occurs as "seventeen thirty" – the number of millivolts registered by the thermocouple; millivolts are the accepted indicator of temperature on this occasion and the lateral traverse of the pen in inches ("eight" in 1755) has to be translated into millivolts using the assumption that the pen traverse (on the x-axis) has been adequately calibrated at 2 millivolts per inch. (Hence the move from "eight" to "sixteen" in 1755.)

We can see that the apparently simple action of reading temperature trades on an entire chain of unquestioned connections. The figures on the digital display are read as a number; this is assumed to denote the amount of voltage across the thermocouple in the oven, the amount of voltage is taken to be an adequate indicator of "temperature", the thermocouple is assumed to be correctly located by the alloy sample, and so on. Although each of these connections is in principle defeasible (that is, the adequacy of the connection can be undermined), assumptions of their adequacy permit short cuts: the numbers read out by T are not merely a hopefully reliable indicator of temperature and so on; Z's comment in line 1756 ("not bad") indicates that, for the immediate purposes of interactional adequacy, T's utterance does indeed articulate the temperature.

The trace is made to speak by virtue of the descriptive work done in the course of the interaction. Frequent reference is made throughout the interaction to the shape and course of the trace. For example:

(C)
1634 Z hmm its going up
1635 T up? – yah – that's going up
1636 Z tee gee
1637 T I don't know if er any I don't know if the relative er
1638 Z yeh I think maybe its er two peaks maybe un
1639 T (few per cent maybe an – few per
1640 cent) but at the last it should be going down

(D)
1718 Z yeah this turned over
1719 T this is the (mark) this is expectation – look – here it comes
1720 (3.0)
1721 Z yeah turned over

(E)
1830 Z its still going
1831 Z ah its slow though – ermm – do that
1832 T it turned a little bit

(F)
1836 B *ahh::::::::::::* its going down – oh it went up a little bit

Let us distinguish between three different analytic positions which might be brought to bear on these data.[5] The first can be called the reflective position. This is the position that descriptions of the character of the trace are determined uniquely by the actual state of the trace. For example, our perception of a bump on the line graph arises by virtue of the *fact* of the bump on the line graph; the bump exists independently of our perception of it. Of course, it is this position which has been roundly criticised in the recent sociology of science literature. Cultural and historical relativism

– the idea that widely differing perceptions of the same phenomenon are commonplace even in science – is invoked to argue that the reflective view is untenable. A second, mediative position, is recommended as an alternative. The argument here is that descriptions of phenomena are not uniquely determined by the phenomena themselves; instead, social, historical, psychological and other factors intercede between the phenomenon and its description. The resulting description is the mediated product of various socio-historical circumstances. In general, much recent sociology of science argument can be characterised as a recommendation that we move from the (naive) reflective position to the mediative position.[6] Frequently, however, it is clear that this second stance itself depends on an implicit realism. By this I mean that one or other description of "what the phenomenon is" appears in the analyst's account as a way of yielding explanatory purchase. A disjunction between alternative descriptions of phenomena is used to justify the introduction of mediating variables as explanans. By contrast, a third constitutive position holds that there is no phenomenon independent of its description. In this view, it is misleading to base analysis on descriptions of phenomena which are assumed unproblematic and the quest for interceding variables is misguided. Instead, the phenomenon is constituted in and through descriptive work, and importantly, this work includes such practices as the assignation of alternative versions, the invocation of relevant mediating circumstances, and so on. Hence, from the constitutive perspective, it is insufficient to treat as incorrigible one or another version of the phenomenon for purposes of explanation: the way in which such versions gain, sustain and lose their corrigibility is the focus of inquiry.

Although this brief treatment of the "problem of descriptions" does scant justice to the complexity of the philosophical issues involved,[7] it nonetheless provides an analytical handle for present purposes. We can apply these three positions to extract (C) as follows. Firstly, we could take the (reflective) position that "up" in 1634 and 1635 is a passive reflection of the actual state of the trace; both Z and T produce "up" because this is the way the trace is. It is easy to see that this argument depends on an assessment of "the actual upness of the trace" which exists prior to and independent of the utterances produced by Z and T. But on what grounds are analysts entitled to evaluate the actual character of the trace? A further difficulty is that apparently similar descriptors occur at different points in the interaction, when participants are confronted by self-evidently different versions of the display. For example, "up" appears as a description of the curve in lines 1634/5, 1836 and 2255 (among many other occasions); "flat" occurs in lines 1648 and 2150. Figure 2 shows the differing state of the trace at these points. Clearly, we cannot assume the same state of "upness" (or "flatness") in each case. It follows that we can do better than treat the documents produced by participants as straightforward reflections of the actual state of the trace.

Secondly, we could take the (mediative) position that participants' perceptions of the trace are mediated by social and historical contingency; in this view, "up" is a product both of the actual shape of the trace and of prevailing social circumstances such as their expectations, their social (and/or cognitive) interests, the constraints of the dominant paradigm, their position in a particular social hierarchy or whatever. This second position attempts to take account of the interpretive flexibility of connections between the actual character of phenomena and the statements which stand as reports upon them. One main difficulty, however, is that this second position again includes an analyst's version of the actual state of the trace. In order to argue the effect of mediating social circumstances, it is necessary to claim that the trace is (or could be) different from the way it was described. In addition, the analyst is required to draw upon features of the participants' environment (or their past history) essentially removed from the immediate scene of the action, in order to concoct ("reveal") the presence and nature of mediating social circumstances.

Thirdly, we could avoid an assessment of the actual character of the trace and instead take the (constitutive) position that the actual state of the trace is the product of participants' accounting; the "up-ness" of the trace is the upshot of the interaction between participants and the document on the chart recorder. On the basis that "up-ness" is created through participants' work, it is interesting to consider how the interaction proceeds as if "up" was a quality intrinsic to the curve, or (to grant more sophistication to our subjects than the naive realism of the reflective position), as if "up" was a matter which could be determined in the face of various intervening social and historical circumstances. In short, analysis from this third perspective asks how participants themselves manage the first and second analytic positions. In other words, what is involved in the practical accomplishment of the reflective and mediative positions such that the shape of the curve achieves its perceived character?

Of course, proponents of the reflective position might rightly argue that the matter is a good deal more complicated than I have indicated. I want to plead for interpretive flexibility on the grounds that identical utterances referred on different occasions to self-evidently different states of the trace. Surely, it could be argued, we should disregard occasions where participants might be merely responding to each other, or to earlier characterisations of the trace. A better test of correspondence would apply when different participants independently produce the same characterisation of the trace. Let us then consider the following example.

(G)
2149 (3.5)
2150 F flat
2151 Z flat? – oh (thats good)
2152 (3.0)
2153 T maybe that is flatter
2154 Z so then maybe it will stay a little bit and then go up

Z's "flat" (2151) is here produced simultaneously with (rather than in response to) T's "flat" (2150). It would be easy to conclude that the shape of the curve gives rise to these two independent but identical assessments of the "fact that it is actually flat". But, let us consider the work that goes into this interpretation. We have to hear the two sounds in 2150 and 2151 as the same word; we have to imbue them with the same meaning; and we have to assume that this meaning relates identically to the same specific phenomenon. At best, the complexity of this work throws doubt on the straightforward conclusion that the utterances result from the independent fact of the trace's shape. In any case, the present data shows a marked difference in inflection between Z and T, thus more readily suggesting the possibility of reading two different social actions in 2150 and 2151 (for example, assertion and question). Even if the inflections and stresses were identical, we would still need to ask "flat in what sense", "flat relative to what" and so on before we could decide whether or not the utterances have identical meaning. The point, then, is not that the simultaneous use of the same descriptor ("flat") establishes that participants both "saw the same thing". Nor would the fact of them "seeing the same thing" entail that the trace was "actually" flat at this point. Instead, we note that "flat" appears to pass as a practically adequate description for participants on this occasion, at least in the sense that there are not subsequent inquiries into "flat in what sense" and so on. My question is: how are subsequent inquiries forestalled or, what makes for this practically adequate sense of "flat"?

Utterances which tell the shape and course of the trace are just one class of documents which, for the most part, pass as practically adequate in the interaction. (We shall consider some instances of disagreement below, where certain references to "what the trace is doing" are themselves subject to scrutiny by participants.) The interactional accomplishment here is the treatment of what is said about the trace as adequately corresponding to the actual state of the trace.

I have suggested that descriptions occurring in the course of the interaction are best understood, neither as direct reflections nor mediated (re)presentations of an independent reality, but as ways of actively constituting the character of the scene which they are taken to be "about". Reference to the current temperature, the shape of the trace and the direction it is taking, all pass as adequate at particular instants in the interaction. This adequacy is achieved through participants' reliance upon chains of unquestioned assumptions. We now need to ask what features of the interaction enable participants to manage the (in principle endless) chain of connections endemic to the interpretation of any document. What features of the organisation of the interaction make possible the accomplishment of descriptive adequacy? In order to tackle this question, let us re-consider extract (C) in more detail.

6. The judicious juxtaposition of documents

Z's observation that the trace is going up (line 1634; see Figure 2) appears to generate a question and self-confirmation from T (1635). Z's following utterance ("tee gee") invokes the symbol for the glass transition temperature (Tg), the point at which amorphous alloys begin their transition to the crystalline state. Hence, Z's utterance (1636) provides for the reading of "up" (1634) as the kind of "up" which both Z and T know to be associated with changes in resistance at the glass transition temperature. This delimits the range of permissible senses of "up", by appeal to a family of events which can be subsumed under the general phenomenon of the glass transition temperature; "tee gee" says, in effect, "it's not any kind of 'up' we have here – not, for example, a vertical step – the way to appreciate the 'up' we are confronting is to take it as a member of the class of 'ups' associated with the glass transition temperature".

It is important to note that "tee gee" works both to modify or elaborate the perception of "up", and acts as a candidate explanation of the "up". The connection between the utterance "up" and the underlying phenomenon "tee gee" is reflexive such that each simultaneously modifies the sense of the other. But the rendering of "tee gee" as a possible explanation for the "up", also provides for the independence of "up" and "tee gee". The thing can be seen to be going up because the oven has reached the glass transition temperature for this particular sample. Now the effect of treating the utterance which provides the sense of the initial phenomenon as a potential explanation for that phenomenon is to background the essentially reflexive character of the initial connection. In other words, to the extent that "tee gee" stands as a potential explanation for the shape of the trace, it assumes an independence of that shape. It is the Tg which gives rise to the upward turn of the trace. The important consequence of "tee gee" acquiring the status of relative independence is that "up" also assumes independence. Although "up" was given its sense by virtue of "tee gee", it now appears as if the "up" all along had its own objective character. The juxtaposition of utterances thus serves both to mutually elaborate their senses and to accomplish their independent objectivity.[8]

We should also note that "tee gee" (1636) engenders subsequent discussion about the possible future course of the trace. Thus lines 1637–1640 raise the issue of whether or not participants can expect a double hump ("to peaks") at the point of glass transition and what kind of rise is to be anticipated ("few per cent maybe") before the trace takes a downward course. The business of "upness" is now left behind in favour of a discussion of the likely next development. The issue of correspondence between description and actual state of the trace, and hence the "fact" of the trace's upward trend, passes (at least for the moment) as interactionally adequate. It is thus clear that "tee gee" fulfils another important function: the invocation of this particular document achieves *interactional closure* on the adequacy of the described state of the trace.

The analysis of extract (C) begins to show how the juxtaposition of documents, in this case utterances, can accomplish the fact of the trace's current shape. But, of course, participants' interpretive work involves much more than their talk about the trace. Let us now consider an example involving the juxtaposition of different kinds of document:

(H)

401 T this one er the temperature coefficient looks a little bit bigger
402 Z it looks like er the copper zirconium two
403 T mmm bigger than that
404 Z nononoalmostthesame
405 T (con)accordingtothewayyouusuallycentrethething(there)
406 Z exactly the same you'll see
407 T but the copper zirconium two (came out here)
408 Z but it looks the same to me
409 T (hnn) ((T walks out of lab.))
410 (14.0)
.
.
.
417 (10.0) ((T brings in CuZr2 chart dated 15–1–81; paper rustling))

418 T now itser one two three – three divisions one two three its three
419 divisions – (maybe four) maybe a little bit more (temperature)
420 coefficient
421 (2.5)
422 T hmm? hmm hmm
423 Z well I think its the same I mean can I start to heat now?
424 T yeh

In this extract, T and Z discuss the slope of the curve at the point where the pen has traversed less than two inches across the chart (along the X-axis; see Figure 2). T's use of "temperature coefficient" (401 and 419–420) indicates how the chart is being read. The trail of ink on the chart recorder is seen as a line with a slope. Although, on other occasions, participants spoke of the "extent of decrease in electrical resistance per degree increase in temperature", at this point both T and Z "see directly" the temperature coefficient on the chart before them. Neither express any misgivings in the use of the ink trail as document of the temperature coefficient, even though at other times queries arose as to whether or not the pen was moving in the right direction (as occurs in this transcript at lines 316–339), whether the thermocouple was working accurately, whether the contacts were affixed correctly and so on. However, in the present instance there is disagreement over the extent of the temperature coefficient. Specifically, the disagreement arises by way of reference to the results of an experiment carried out seven days earlier on a different composition of the amorphous alloy (CuZr2). Note here how the utterance "copper zirconium two" (402) passes as an adequate reference to relevant aspects of this earlier result. (It does not, for example, work as reference to a chunk of the metal alloy, nor to any other experimental treatment of the material.) Given T's characterisation of the present slope in terms of "temperature coefficient", it is this feature of "copper zirconium two" which is heard as being commented upon by Z. The interaction in lines 401 and 402 thus elaborates relevant particulars of both present and past slopes. At the same time, the "fact" of correspondence between ink trace and temperature coefficient is solidified. The interaction moves into a comparison between temperature coefficients and thus achieves closure on the adequacy of the slope as an indicator of temperature coefficient.

The sense of the present slope (Cu50Zr50) is given by reference to the state of a past slope (CuZr2). This comparison nicely illustrates the reflexive tie (in Garfinkel's sense of the term) between "present temperature coefficient" and "the state of the past slope". The character of the present slope depends, at this instant, on the actual state of the past slope. Of course, the matter could easily be settled if we could fix the character of both slopes: if the actual state of both slopes was known, then the question of their relative size would be a cinch. There is, however, an interesting asymmetry in the kind of documents invoked in this discussion. The slope (of Cu50Zr50) is indexed both by participants' utterances and the pen trace before them. Until line 417, the slope of CuZr2 is merely talked about, that is, participants' utterances are the only documents of the state of the CuZr2 slope. Lines 401–409 can be understood as an argument about the character of CuZr2 slope in which both participants perform mediative analysis. Their concern is to specify the presence or absence of distorting factors which might be intervening between "the actual slope of CuZr2" and mere recollections as expressed in verbal (re)description. T's utterance (405) indicates that he initially treats Z's insistence on the similarity of the slopes as an artefact of Z, failing to take into account the way in which the pen is usually centred (that is, what Z perceives as a similarity is in fact a difference, once the method of centring has been taken into account). But this appeal to a contingent aspect of the experimental practice is unsuccessful (406) and T reverts to a redescription of the CuZr2 chart (407). When this further attempt to use utterances as an adequate document also fails, T finally invokes a quite different kind of document viz. the chart of CuZr2 which he fetches from an adjoining office. Even the presence of this chart is supplemented with T's utterances in 418–420. T's counting of the squares on the graph paper provides a caption for telling what the chart before them "shows".

For Z, who maintains that the slopes are similar, the disagreement appears to hinge on the indeterminancy of the present state of the Cu50Zr50 trace rather than on the inadequacy of utterances as accurate documents of the CuZr2 result. (It is Z who first invokes "copper zirconium two" as a point of comparison.) The state of the present trace is not yet fixed, in Z's view, so that assessments of its slope currently depend on individual evaluation ("to me" – 408), on perception rather than on the nature of the trace ("it looks like" – 402; "it looks the same" – 408; "I think its the same" – 423) and on their being made in advance of the state of affairs as it may eventually reveal itself ("you'll see" – 406). Note that it is precisely Z's reference to himself as an agent of perception ("it looks the same to me" – 408) that T takes as indicative of a possibly unreliable connection between utterances about CuZr2 and its actual character.

In this example, T's claim that the trace exhibits a larger temperature coefficient is buttressed by the use of a document (chart of CuZr2; 15–1–81) as a definitive record of a past achievement. This appears designed to close off the possibility of inaccurate recall, but it fails to adduce any obvious token of agreement from Z. The asymmetry in comparison between mere utterances (about CuZr2) and the actual chart before them (Cu50Zr50) is rectified, but Z claims to see no differences between two charts before him ("well I think it's the same"), despite T's use of captions to direct him to read a difference (lines 417–420). There does not seem a lot more T can do at this point; having effected the juxtaposition of two documents of (claimably) equal interpretive reliability, he has played a major card and Z is still unyielding. T accepts Z's attempt to close off the topic ("can I start to heat now?") and attends to the task of monitoring the correct operation of the heat supply. The offer and acceptance of attention to this new task achieves interactional closure on the issue of similarity or difference between present and past output traces.

(I)

901 T hmm no sign of oxidation uhh
902 Z but this one has been oxidized maybe as much
903 as it could maybe oh yeh you want yes=
904 T no:::: no no ::::
905 T =oxidized will change colour
906 Z no wait a sec yeah okay but – where is the (millec)

907 did we throw it ah here it is there – okay oh this
908 is not a good one it doesnt show oxidation does this
909 show it ((paper rustling 2.5)) and jus around here there
910 is oxidation maybe
911 T yeah yeh yeh

Whereas extract (H) showed participants "seeing" the temperature coefficient, the above example shows that the trail of ink is made to work for the presence of "oxidation". Briefly, participants trade on the assumption that oxygen contaminating the immediate environment of the heated alloy sample will cause a spurious increase in its electrical resistance. T's utterance (901) portrays the line as a document of the absence of such an increase. Z invokes for the sample a past history, the idea that the sample has already experienced maximum oxidation some time during the initial phase of the experimental run (902–903). Z does not take direct issue with T's proposal that the shape of the trace documents an absence of oxidation. Instead, Z attempts to provide an explanation for this absence.[9] T clearly objects (904) to Z's alternative proposal by drawing attention to the colour of the sample in an attempt to undermine Z's interpretation. The darkening of an alloy sample is characteristically indicative of oxidation. At line 905, Z attempts to deal both with T's description of the state of the trace (901) and with his invocation of the lack of colour change (905). Z's immediate problem is that he objects neither to the idea that there is no oxidation, nor to the rule that oxidation is indicated by a colour change. So Z does not counter T's proposal by challenging the connection between T's statements and "the actual facts of the matter". Instead, Z introduces a quite separate kind of document into the discussion, namely, two output charts lying on a nearby table. The first of these appears unsatisfactory (907–908), but the second is said to reveal oxidation occurring at an earlier stage of heating (909–910). At least for the moment, T appears to accept this latter document as indicating the possibility of prior oxidation of the present sample ("yeh yeh yeh" – 911).

In many similar examples a claim (that the temperature coefficient is bigger than previously; that there is no sign of oxidation) is rendered problematic by construing one set of utterances as "mere talk about the state of affairs" rather than as "things visible in the documents before us". Although attempts are made to deal with this situation by the subsequent invocation of a different kind of document (viz. the introduction of charts as argumentative props), this does not in itself always settle the issue. The outcome appears quite different in the cases in extracts H and I. This suggests that whether the juxtaposed documents are visible charts or speech utterances does not *determine* the outcome of a disagreement about the adequacy of description. It is clear, however, that in every case documents are used to appeal to a past state of affairs, by reference to which the current state of the trace is presumed knowable. Documents are thus used to fix a presently ambiguous state of affairs by tying it to (what is presented as) a previously settled objectivity.

The appeal to a past (settled) state of affairs can be somewhat tenuous when manifested by a single document. A lone document can be said to be unrepresentative of the past; its presentation as a *single* document can invite the reopening of questions about its character. For example, how can we be sure that yesterday's chart does in fact show what we (then) thought it did. The more successful strategy requires the invocation of a document as one *part* of an established state of affairs. The document is then introduced as merely representative (speaking on behalf) of an unmentioned number of similar other documents. This kind of strategy is most evident in participants' comments about the overall shape and course of the shape:

(J)

546 T IF YOU ARE NOT OVERheh – okay now we leave to do this
547 Z now its not necessary to watch it
548 (1.5)
549 Z not really I think its the same thing
550 T (I prepare now for the talk)
551 Z you you you think this will hold up?
552 T yeah yeah yeh
553 (3.0)

(K)
855 (2.5)
856 J well that one is a very thrilling thing it
857 goes down in eactly the same way as the others
858 Z this one? yeah:: thats okay

(L)
1425 T nyarhh
1426 J its sort of boring watching this thing do nothing isnt it

(M)
1433 Z you might as well have drawn the thing you know
1434 J the line is straight ((yawn))
1435 Z and forget about it
1436 T narh thas true hahah

Each of these extracts includes attempts to formulate the overall character of the current state of the trace. Participants are here concerned with what, after all, the trace is turning out to be. In both (J) and (K), the state of the trace is compared with "others", either explicitly (857) or implicitly (549). In a similar way, descriptions of the merely routine character of the trace in (K) and (L) constitute the state of the trace in terms of unspecified other traces which have preceded it. Clearly, hearers of "its sort of boring watching this thing do nothing" (1426) understand a very specific kind of "nothing". In another sense, it is difficult to conclude that the trace is doing nothing, given participants' marked concern over whether the trace is moving in the right direction (316–339); whether its electrical contacts are well fixed (343–346); if its temperature coefficient is larger than a previous sample (extract (H) and 923–926); whether it is showing signs of oxidation (901–922), and so on. Although lively throughout the rest of the interaction, in these extracts such issues are treated as resolved and unremarkable for the immediate purposes of formulating the overall character of the trace. "Nothing" (1426) claims a similarity with, or lack of deviation from, other traces whose common character is presumed. The present trace is doing nothing in the sense that it conforms to this common feature.

In all these instances, then, the character of the trace is achieved by claiming its membership of a family of similar traces. The "others" are portrayed as a collectivity with certain common features. The way in which "the thing goes down" (857) is one such property common to members of the invoked collectivity and it is this property which is used to establish the family membership of the present trace. The property common to the family (whether this is "the way the thing goes down" or the way it does "nothing") features as an established and incontestable aspect of various other traces by virtue of an unmentioned edifice of similarity relations between existing members of the family. In addition, the objectivity of this membership characteristic is based on unspecified past achievements, the sense of which is presumed settled by virtue of unmentioned prior actions.

7. Discussion

It would be premature to adduce a set of rules for the interactionally adequate use of descriptions in experimental science on the basis of the few short analyses presented here. Nonetheless, it is worth restating some of the central observations of these analyses so as to highlight their common themes. I began by showing how the perceived quality of the trace is achieved by descriptive work which makes out that, for the immediate practical concerns of the interaction, this quality exists independently of, and prior to, the work of description. I noted in particular the importance of the juxtaposition of documents for achieving descriptive adequacy. Participants' invocation of a document, its juxtaposition with a prior document, can elaborate the sense and meaning of the visual display before them. In addition, the juxtaposed document can help accomplish an objectivity for the display to which the description refers. Further, the introduction of a juxtaposed document shapes the course of the interaction; potential concerns with the adequacy of prior descriptions are to some extent backgrounded by participants' attention to issues raised by the new document. In this way, the juxtaposed document contributes to the achievement of interactional closure on descriptive adequacy.

We noted that the way in which documents were introduced into the interaction was more significant than their material character (for example, whether the document was a chart rather than a speech utterance). As became clear in the final extracts (J, K, L, and M), formulations of the overall character of the visual display involve claims about the trace's membership of a similar collectivity of past displays. Reconsidering the earlier analyses, we can see that juxtaposed documents are introduced so as to claim a similar sense of family membership. The invocation of Tg suggests the present state of the trace can be made sensible in term of its membership in a class of phenomena (the changing structural characteristics of metal alloys at the glass transition temperature) which comprises past events sharing similar characteristics. The production of charts resulting from previous experimental runs appeals to the possibility that current states are compatible with a collectivity of established results. Repeatedly, then, attempts to render the present state sensible draw upon representations of the past. Documents are introduced as manifestations of the established past and juxtaposed with the present phenomenon so as to give it meaning. As we have seen, these representations conjure up sets of past achievements which are taken to be both similar and incorrigible. The achievements are construed as similar because they are referred to in terms of family characteristics; they pass as incorrigible by assuming their pastness is sufficient grounds not to doubt the legitimate family membership of any one achievement.

Clearly, this kind of analysis only scratches the surface of issues of historicity involved in experimental description. Whereas this paper has concentrated on isolated fragments of the interaction, an obvious next task is to attempt an examination of the interaction as a whole. In particular, this will enable us to consider more carefully what is peculiar about an interaction that is tied to a visual display which is unfolding as the interaction goes on. How exactly do the direction of the interaction and the drift of the trace mutually organise one another?

Notes

1. At least two key questions require consideration. First, what is the

analytic status of these methodological horrors? Although the perspective outlined above might appear to claim a privileged ontological status for the principled fallibility of practical actions, no such claim is intended. The assertion of fallibility appears here as an analyst's gambit, a methodological heurstic; this backdrop of putative fallibility enables us to highlight the interpretive work of our subjects. This is not to claim, for example, that any particular scientific inference is in fact impossible, nor to suggest that any particular interpretive manoeuvre is un-warranted. The present argument has no pretensions to this kind of evaluation of scientific work. But the notion of methodological horrors does acknowledge the possibility of a form of philosophical idealism applied relentlessly to each and every instance of interpretation. The analysis is not an attempt to demonstrate the ultimate fallibility of interpretive action, but an exploration of the potency of epistemological scepticism. Second, for whom are the methodological horrors horrible? If they exist at all, can we reasonably proceed on the assumption that they could or should bother practicing scientists? Evidence from existing studies of scientific work indicates that scientists are not mindful of the full horror of methodological fallibility. On the other hand, they do appear mindful of the possibility of technical errors. At least in their public and post hoc reconstructions scientist pay careful attention to the methodological correctness of their interpretations. The attention given these technical errors can be regarded as the strategic navigation of the tip of an epistemological iceberg. Potentially threatening horrors are reduced to matters of merely technical (and therefore correctable) difficulty. Mistakes, errors and red herrings are thus prevented from threatening the explanatory enterprise itself because they are rendered merely peripheral to the interpretive process.

2. Several factors, notably the unobtrusive size of the recorder and the fact that by this stage of the study I had been routinely using the tape recorder for several months (since the end of October 1980), make it reasonable to assume that at least some portions of recorded interaction were relatively unaffected by the presence of the tape recorder. In any case, I suggest my particular analysis of the talk is not greatly affected by whether or not participants were aware of the tape recorder. In a subsequent discussion, three participants volunteered that they usually forgot about the tape recorder soon after it was introduced.
3. Asterisked line numbers denote subsequent reconstruction of the state of the trace when transcribed utterances took place.
4. The sensitivity of this indexing system can be gauged by noting that the run proceeded at a heating rate corresponding to a horizontal pen movement of about one inch every few minutes. At the time of Z's comment, "well it should turn now" (1935), each line of transcription takes up an average (due in part to pauses in the talk) of about 3 seconds. But it took approximately 10–15 seconds to note down the pen position corresponding to an utterance. Although it was quite easy to

assess the position of the relatively slow moving pen, the recorded position is thus subject to an error plus or minus approximately 3 lines of transcription. However, this error is not entirely random, since the serial order of utterances (and their corresponding positions on the trace) is preserved by the tape recording.

5. A detailed exposition of these different positions is given in Woolgar (1983).
6. For example, Barnes (1978) speaks of supplanting the passive contemplative view of knowledge with a more active sociological view; Mulkay (1979) speaks of revising the standard view of science in favour of a position which does not regard knowledge as stemming solely from the objective character of the natural world. See also discussion in Woolgar (1983).
7. Or perhaps it is philosophical discussions which do scant justice to the complexity of the practical problems of description!
8. In effect, this entire strategy amounts to a "micro" application of the "splitting-inversion" model of scientific discovery. See Latour and Woolgar (1979).
9. Ethnographically, it is possible to argue that the discussion here took place in the context of a long series of attempts to design ovens free from contaminating oxygen. T's remark (901) can thus be heard as a claim for the apparent success of the present experimental arrangement.

References

Altounian, Z., Tu Guo-hua, and Strom-Olsen, J.O. (1981). *The crystallization of amorphous CuZr2.* Laboratory report. July.

Barnes, B. (1977). *Interests and the growth of knowledge.* London: Routledge and Kegan Paul.

Chaudhari, P., Giessen, B.C., and Trunbull, D. (1980). Metallic glasses. *Scientific American* 242:98–115.

Garfinkel, H. (1967). *Studies in ethnomethodology.* Englewood Cliffs, NJ: Prentice Hall.

Garfinkel, H., Lynch, M., and Livingston, E. (1981). The work of a discovering science construed with materials from the optically discovered pulsar. *Philosophy of the Social Sciences* 11:131–158.

Latour, B., and Woolgar, S. (1979). *Laboratory life: The social construction of scientific facts.* Beverly Hills: Sage.

Lynch, M. (1982). Technical work and critical inquiry: Investigations in a scientific laboratory. *Social Studies of Science* 12:499–533.

de Saussure, F. (1959). *Course in General Linguistics*, trans. Wade Baskin. New York: Philosophical Library.

Schenkein, J. (1978). *Studies in the organization of conversational interaction.* New York: Academic Press.

Sturrock, J., Ed. (1979). *Structuralism and since: From Levi-Strauss to Derrida.* Oxford: Oxford University Press.

Woolgar, S. (1983). Irony in the social study of science. In K.D. Knorr-Cetina and M. Mulkay (Eds.), *Science observed: Perspectives on the social study of science*, 239–266. London: Sage.

The externalized retina: Selection and mathematization in the visual documentation of objects in the life sciences

MICHAEL LYNCH
Department of Sociology, Boston University, Boston, MA 02215, USA

1. Introduction

This study examines visual displays used in scientific publications. Two collections of displays are discussed in reference to two themes on the constitution of the scientific object: "selection" and "mathematization." Selection concerns the way scientific methods of visualization simplify and schematize objects of study. Mathematization concerns how such methods attribute mathematical order to natural objects. I do not mean just to use the illustrations simply as *examples* of the themes, but also to critically examine the themes in light of the analytic details of the visual displays.

Relatively few science studies directly deal with the concrete details of visual displays. Verbal propositions, arguments, references, analogies, metaphors, and 'ideas' have received much greater attention as constituents of scientific reasoning and rhetoric. This imbalance may be due to the fact that methods for analyzing verbal materials are more developed than those for analyzing pictures. The fact that writing is the dominant medium of academic discourse is not incidental; while pictorial subject matter is alien to written discourse, and requires a *reduction* to make it amenable to analysis, written subject matter can be iterated without any 'gap' within the textual surface that analyzes it. Nevertheless, visual displays are distinctively involved in scientific communication and in the very "construction" of scientific facts.[1]

This study is based on the premise that visual displays are more than a simple matter of supplying pictorial illustrations for scientific texts. They are essential to how scientific objects and orderly relationships are revealed and made analyzable. To appreciate this, we first need to wrest the idea of *representation* from an individualistic cognitive foundation, and to replace a preoccupation with images on the retina (or, alternatively 'mental images' or 'pictorial ideas') with a focus on the 'externalized retina' of the graphic and instrumental fields upon which the scientific image is impressed and circulated.[2] For sociological purposes, the "real" object is the representation in hand, e.g., the visual display, and not the invisible phenomenon or abstract relationship "out there."[3] Furthermore, we need to recall that visual documents are used at all stages of scientific research. A series of representations or renderings is produced, transferred, and modified as research proceeds from initial observation to final publication. At any stage in such a production, such representations constitute the physiognomy of the object of the research.

To show how scientific documents are revelatory objects, and, moreover, objects which simultaneously analyze what they reveal, I have put together two collections of illustrations from various scientific texts. The specific cases are not selected to sample common types of illustration; instead they permit an examination of transformational properties which are less evident in other, perhaps more typical, illustrations.

The first collection consists of 'split-screen' juxtapositions of photographs, diagrams, and sometimes 'models,' each of which proposes to represent 'the same thing.' The split-screen format enables us to discuss procedures of selection or simplification by examining how diagrams transform photographic depictions. The second collection consists of illustrations which display how a natural terrain is turned into a graphic field. These illustrations will be analyzed for how they identify substantive properties of depicted 'objects' with the mathematical parameters of a graph.

Although I shall focus on published illustrations, I will use the illustrations as an occasion to discuss the research process. Published illustrations are not self-sufficient descriptions of research processes.[4] They do not directly reveal how researchers produce and utilize visual documents during laboratory projects. They are displayed to specific audiences, and any adequate interpretation of them requires an understanding of the written text of the articles in which they appear, which, in turn, requires a technical expertise. I have chosen examples from biology, particularly from analytic microscopy, and, in one case, field biology, because they are more familiar to me than other scientific fields. My understanding of the production of the particular illustrations is nonetheless very limited, though it is sufficient for my purposes here.

The procedures of selection and mathematization that I discuss here are not simply 'done by' illustrations. Illustrations instead provide a literary convenience for the purpose of succinctly exhibiting those constitutive practices. The way documents are used in natural science texts thus converges with their analytic use here; my discussion parasitizes properties of the documents it analyzes. Foremost among these parasitized properties is that of how the illustration is an autonomous surface that is nonetheless contained within a text, and which can be used in various ways by the text's discourse to invite a reader to 'see what is being said.'[5] I will rely upon the availability of illustrations for the devices of literary revelation and exemplification in order to introduce a reading of the illustrations which differs from that used in natural science research. Instead of using pictures as evidence for naturalistic claims about objective entities or relationships, I will use them as evidence of methodic practices, accomplished by researchers working together in groups, which transform previously hidden phenomena into visual displays for consensual 'seeing' and 'knowing.' For a naturalistic fascination with what pictures show, these transformations are largely taken for granted, and yet are relied upon as contextual elements of a scientific object's sensual availability. I will discuss only a limited array of such contextual features, while relying upon others for my own textual demonstration. As stated above, the 'adequacy' of what I will say will depend largely upon a use of the documents I analyze to allow readers to independently see what I am saying.

2. Selective perception

In pragmatist, phenomenological, and interpretive sociological traditions, perception is often likened to a filter which selects from, simplifies, and orders an initially chaotic world in terms of the perceiver's projects and interests. Originally, this image was used to describe the operations of the individual "mind" or "consciousness" faced with physical stimuli.[6]

> Human mind, and also the sense-organs and even all organic life, is a 'selective agency.' Owing to their selective activity, the sense-organs filter the physical stimuli by which they are excited. By a subsequent selection, those sensations which serve as signs of things are shifted from the totality of experienced sensations. Selection is responsible for the constancy-phenomena: rhythm into a monotonous succession of sonorous strokes, it groups dispersed dots into rows, figures, and constellations. Whatever *organization* may be found in experience is *bestowed* upon it by the mind working on the 'primordial chaos of sensation.' (Gurwitsch, 1964:28)

In more recent sociological usage, the idea of selection applies to the coordinated practices of groups of people instead of to the psychology of the isolated individual. Scientific research teams are described as agencies of mediation between an uncertain and chaotic research domain and the schematic and simplified products of research that appear in publications.[7] Research teams use laboratory practices to transform invisible or unanalyzed specimens into visually examined, coded, measured, graphically analyzed, and publically presented data. Such ordering of data is not solely contained 'in perception,' but is also a social process – an "assembly line" resulting in public access to new structures wrested out of obscurity or chaos. Instruments, graphic inscriptions, and interactional processes take the place of 'mind' as the filter, serving to reduce phenomena of study into manageable data.

The image of filtering (of selective attention and retention of simplified research products) has much to recommend it. As a sensitizing concept, it is preferable to any account that treats published scientific data as no more than a 'rational reflection' of an independent empirical world. It encourages a sociological interest in research processes and their resultant 'facts,' and an understanding of them as instances of social agency. As such, they have a more general significance than as empirically adequate practices governed by methodological prescriptions.

The idea of selectivity or simplification serves adequately as a starting point, as a sensitizing notion that motivates a sociological interest in how scientific research constitutes objects of study. I will claim here, however, that the metaphor of filtering fails to address significant aspects of research practice. To support that claim, and, more importantly, to point to characteristics of the research process that are missed by the notion of selectivity, I will discuss a series of figures chosen from various scientific texts.

Each of these illustrations displays a similar format. I will use the illustrations to argue that the notion of simplification is too simple, and that it glosses over features of transformational practices which, when examined in their own right, do not seem to be matters only of filtering or selection. While a few published illustrations will not give access to the lively complexities of laboratory work, and of the *in situ* work of transforming specimens into "facts," they are adequate for the purpose of reexamining the idea of selection or simplification in scientific 'perception.'

Figures 1 and 2 each exhibits a similar 'split-screen' format. A photograph is paired with a diagram, both of which purport to represent the same phenomenon. The diagram is a 'rendering' or transformation of the photograph, and not vice-versa. The relationship between the two initially appears to be governed by the diagram's selection of information to compose a simplified presentation more closely aligned with the purposes of textual presentation. Close examination, however, reveals that the transformational process consists of selection and simplification only in part. The process synthesizes form, as well. Most importantly, it strives to identify *in* the particular specimen under study 'universal' properties which 'solidify' the object in reference to the current state of the discipline. The latter claim will be explicated in a discussion of Figures 3–5, which provide an interesting variation on the split-screen imagery of the earlier figures.

2.1 A 'directional' relationship between paired representations

Both figures have a similar format: a photograph is placed alongside a drawing of similar size and orientation. One can easily see

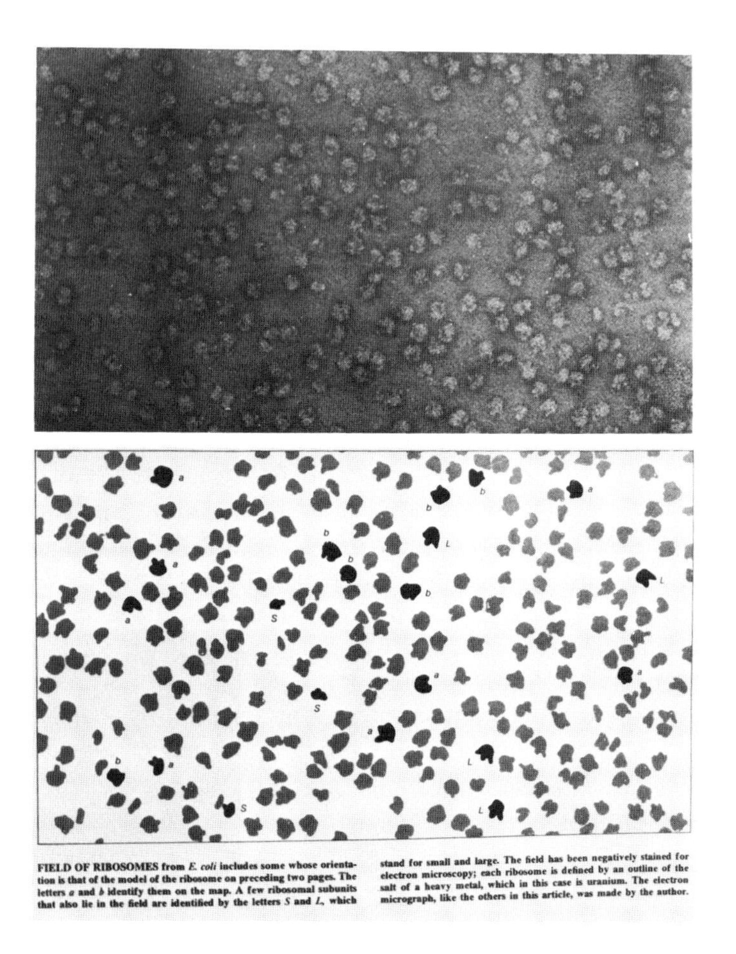

FIELD OF RIBOSOMES from *E. coli* includes some whose orientation is that of the model of the ribosome on preceding two pages. The letters *a* and *b* identify them on the map. A few ribosomal subunits that also lie in the field are identified by the letters *S* and *L*, which stand for small and large. The field has been negatively stained for electron microscopy; each ribosome is defined by an outline of the salt of a heavy metal, which in this case is uranium. The electron micrograph, like the others in this article, was made by the author.

Figure 1. Field of ribosomes

Figure and caption from James A. Lake, "The Ribosome," *Scientific American* 245.2 (August, 1981), p. 86. W.H. Freeman and Company, NY. Reprinted with permission

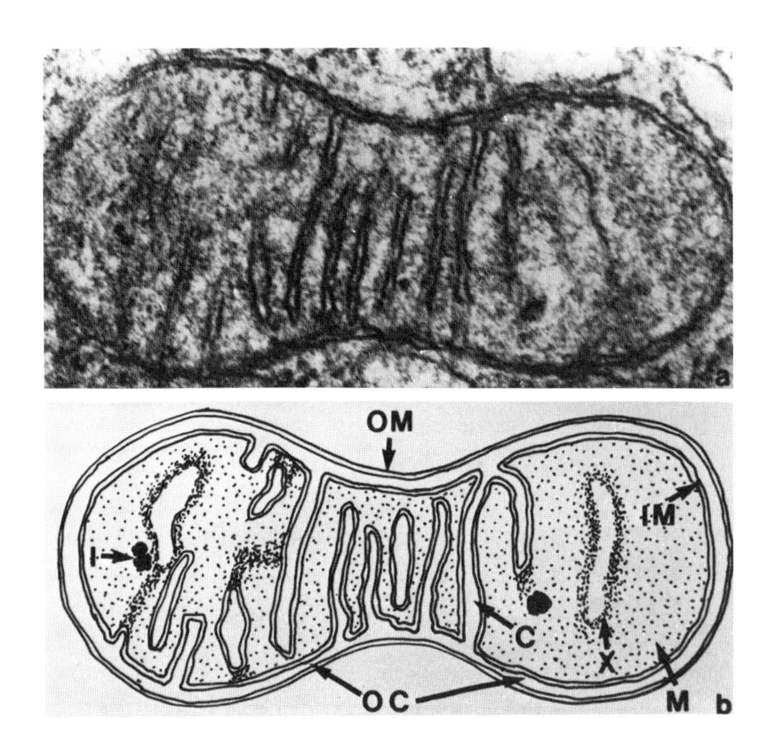

Figure 2. Mitochondrion

Figure and caption from L.T. Threadgold, *The Ultrastructure of the Animal Cell*, Second Edition, p. 321. Oxford: Pergamon Press, 1976. Reprinted with permission

(a) Mitochondrion from the pancreas of the mouse showing all the main characteristic features of this organoid which are labelled in the diagram in (b). OM and IM, outer and inner membranes; I, intramitochondrial granule; C, cristae; X, oblique section of cristae; OC, outer compartment; M, matrix.

that both are presented as "images of the same thing," only they are different. The diagram is a schematic representation of what can be seen in the photograph. The members of the pair have a *directional* relationship to one another: each is an independent representation, but they are not equivalent. One depends upon the other: the diagram operates upon what is shown in the photograph (unlike, for instance, two different technical renderings of a same thing,[8] such as can be seen in comparisons between x-ray and optical photographs of a distant galaxy or nebula). Although the diagram can be seen as a schematic version of the photograph, the photograph is not to be taken as a schematic representation of the diagram. The pair thus shows a sequential ordering; the photograph being an "original" and the diagram a rendering of it.[9] This might allow us to say that the diagram is a "reduction" of the photograph, a simplification that is more congruent with the didactic or representational purposes of the respective text in which it appears. Relative to the diagram, the photograph appears to be more "original" material," whereas the diagram is more evidently analyzed, labeled, and 'idealized.'

The paired relationship is not only sequential; each photograph is not simply more original 'in time,' but is presented relative to the diagram as *original evidence*. The photo's photo-chemical transfers can be invoked to explain its image as something 'more real' than an artistic creation. It is as though the image is imparted by the object itself. Of course, this is not completely so. The photographed materials (i.e., the cellular matrices from the specimen) have been extensively handled in order to prepare them for the picture. Relative to the juxtaposed diagram, however, the photograph less obviously shows the 'hand' of the artist.

2.2 Sequential transformations in photo-diagram pairs

Examination of the figures reveals several kinds of transformation. These include what I have called "filtering" (or selection/ simplification), but they also include transformations which change the characteristics of the photographic images in ways other than reducing information. Several transformative practices are identified below under the rubrics of "filtering," "uniforming," "upgrading," and "defining."

2.2.1 Filtering

The diagrams exhibit a limited range of visible qualities in comparison to the photographs. Unused visibility is simply discarded out of the picture. In Figure 1 the diagram shows a blank, completely empty, white background between the isolated "ribosomes," instead of the grainy and somewhat variegated texture of the photograph.

2.2.2 Uniforming

Visual conventions using color fields, textured spots or cross-hatches, and uniform shading transform variegated fields in the photographs to relatively uniform fields in the diagrams. The visual variation between 'ribosome' profiles in Figure 1 is reduced in the transformation between photo and diagram. In Figure 2, the interstitial spots in the diagram are relatively uniform in size, shape, and distribution. Their uniformity is not 'perfect,' but compared to the photograph the diagram shows less variation in shading and texture.

2.2.3 Upgrading

The sensual qualities of displayed entities are made more congruent with the identities assigned to those entities. Borders are made clear and distinct, lines of uniform thickness are drawn where fragmentary 'lines' or no lines at all were visible in the photographs. Shapes, and divisions between distinct surfaces are made more definite. Dim differences become clear differences of structure, and identifying features are more clearly distinguished against their backgrounds.

2.2.4 Defining

Entities are not only made more like one another, they are more clearly distinguished from *unlike* entities. Sensual qualities of the image work in concert with linguistic labels and pointers to code and categorize entities. What counts as likeness or distinctness depends upon the analytic purposes of the text in which the figures appear. In Figure 1, a selection of ribosome profiles is distinguished from others by being darkened, and then labeled with letters. The caption distinguishes these in terms of qualities of a model displayed elsewhere in the text. The labels "code" the profiles for size and orientation. In this case, the darkening is not an enhancement of distinguishable color or texture differences but a way of singling out representatives of categorical types based on other grounds.

In Figure 2, the lines and textures of the diagram adjust the sensual qualities of what is shown to more clearly 'respond' to the labels connected by the pointers. The figure-ground relations, initially visible but less apparent in the photograph, are *skewed* in reference to the ostensive definition: the word points to the entity, and the entity responds *as a literary accomplishment* by leaping out from its background.

The juxtaposition of photograph and diagram does more than show how diagrams reduce and add to the information from the photographs. It shows how the diagrammatic renderings, and the textual usage they support, are not fashioned from out of 'pure' chaos, but as a serial transformation of more original, though 'messier,' renderings. The photographically blurry ribosomes that Figure 1 clarifies can be seen to 'support' a diagrammatic deletion of irregularity, overlapping, and discontinuity in the original. It could be said that the photograph is an imperfect representation of an actual array of ribosomes, and that the diagram represents this actual array more faithfully.

I do not wish to dispute the above arguments, although there have been historical cases where the diagrammatic completion of situated visual experience has been responsible for some hilarious monsters,[10] and in ordinary lab research less dramatic artifacts are quite common (Lynch, 1985b:81).

What I wish to argue is that, *relative to the photograph*, the diagram is an *eidetic* image[11] and not merely a simplified image. I emphasize this relativity because it could also be argued that relative to what would be seen in the microscope the photograph is an eidetic image, or that relative to the original specimen the stained and otherwise prepared slide in the microscope is an eidetic image (note how the caption in Figure 1 states that the ribosome in the micrograph is "defined by an outline of the salt of a heavy metal"). By "eidetic image" I mean an icon of what Heidegger (1967:101–102) calls "the mathematical" in the sense of *mathesis universalis*: the theoretical domain of pure structure and universal laws which a Galilean science treats as the foundation of order in the sensory world. When paired together, such as in Figures 1 and 2, the differences between photograph and diagram can be resolved by associating the photograph with the unique, situationally specific, perspectival, instantaneous, and particular aspects of the thing under examination while the diagram brings into relief the essential, synthetic, constant, veridical, and universally present aspects of the thing 'itself.' The duality is not absolute; the diagram is not simply an "ideal" image while the photograph is "empirical." Both photograph and diagram exist on a common textual surface, and as such depend upon the artifices of inscription and interpretation while representing some worldly object. It is only when we compare the two that we can see how the diagram leans slightly more in the 'eidetic' direction than does the photograph.

There is also a cumulative element: a ribosome is given a label and made to stand out against its background in part because it can be argued that it is one like innumerable others. The particular vagaries of *this* specimen are set off against the commonalities of an indefinite series of similar specimens in the context of an established field which credits the existence and defines the characteristics of ribosomes in general. The conventions of representation are thus more than artistic devices, they take their authority from previous experience and the state of the scientific field to competently build on a body of assumptions about the represented structures.

The image of the filter is inadequate. The Jamesian notion is dichotomous. It implies that outside the filter is a structureless chaos, inside organization. The figures above show more of a continuum of representations modifying the products of previous observations and representations. Each of these representations selects from a prior representation, while exhibiting a dependency on pre-established formations visible in the prior and at the same time 'upgrading' the orderliness and utility of those formations. Order is not simply constituted, it is *exposed, seized upon, clarified, extended, coded, compared, measured, and subjected to mathematical operations*. (These latter qualities will be examined in the discussion of "mathematization," below.) These terms, these modifications, depend upon a prior, though relatively indeterminate, *something* which is successively modified into a more 'useful,' and at the same time 'theoretical' object. To further exemplify and build upon the later theme, we will examine a variant of diagrams commonly called "models."

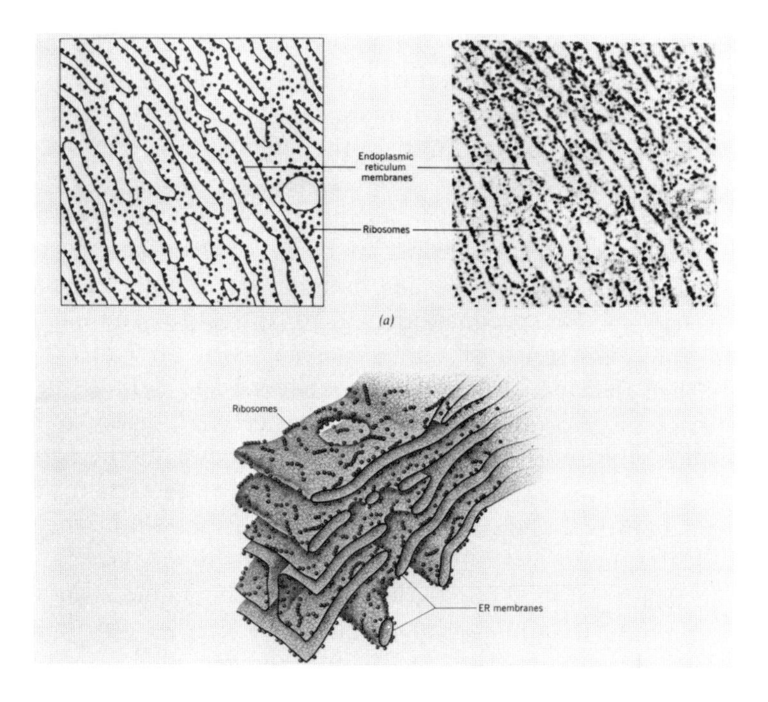

Figure 3. Endoplasmic reticulum

Figure and caption from Grover C. Stephens and Barbara Best North, *Biology*, p. 79. New York: John Wiley and Sons, 1974.

Endoplasmic reticulum and ribosomes. (a) Rough ER is usually found in cells that are actively engaged in protein synthesis, such as this cell from the pancreas of the rat. (b) Drawing shows the three-dimensional structure of rough ER.

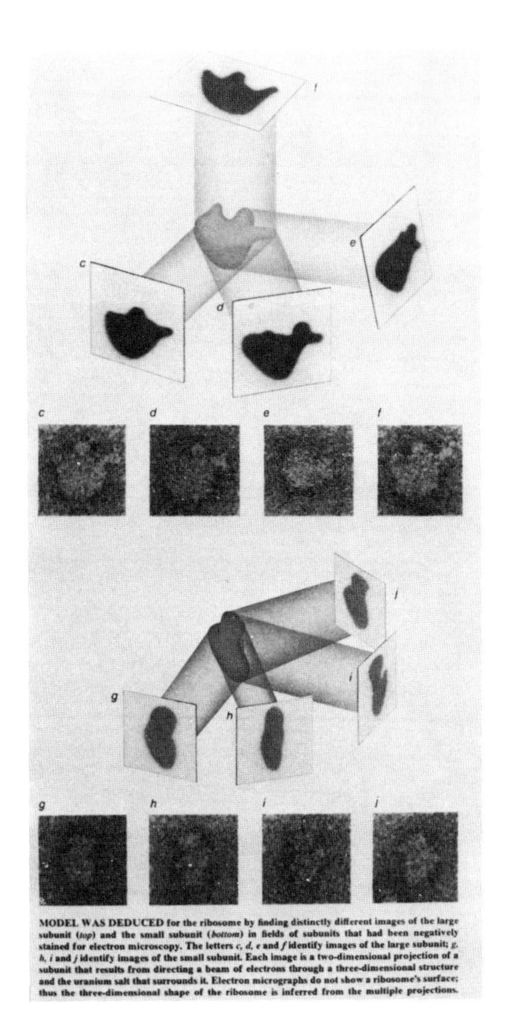

MODEL WAS DEDUCED for the ribosome by finding distinctly different images of the large subunit (*top*) and the small subunit (*bottom*) in fields of subunits that had been negatively stained for electron microscopy. The letters *c*, *d*, *e* and *f* identify images of the large subunit; *g*, *h*, *i* and *j* identify images of the small subunit. Each image is a two-dimensional projection of a subunit that results from directing a beam of electrons through a three-dimensional structure and the uranium salt that surrounds it. Electron micrographs do not show a ribosome's surface; thus the three-dimensional shape of the ribosome is inferred from the multiple projections.

Figure 4. Ribosome model

Figure and caption from James A. Lake, "The Ribosome," *Scientific American* 245.2 (August, 1982), p. 87. W.H. Freeman and Company, NY. Reprinted with permission

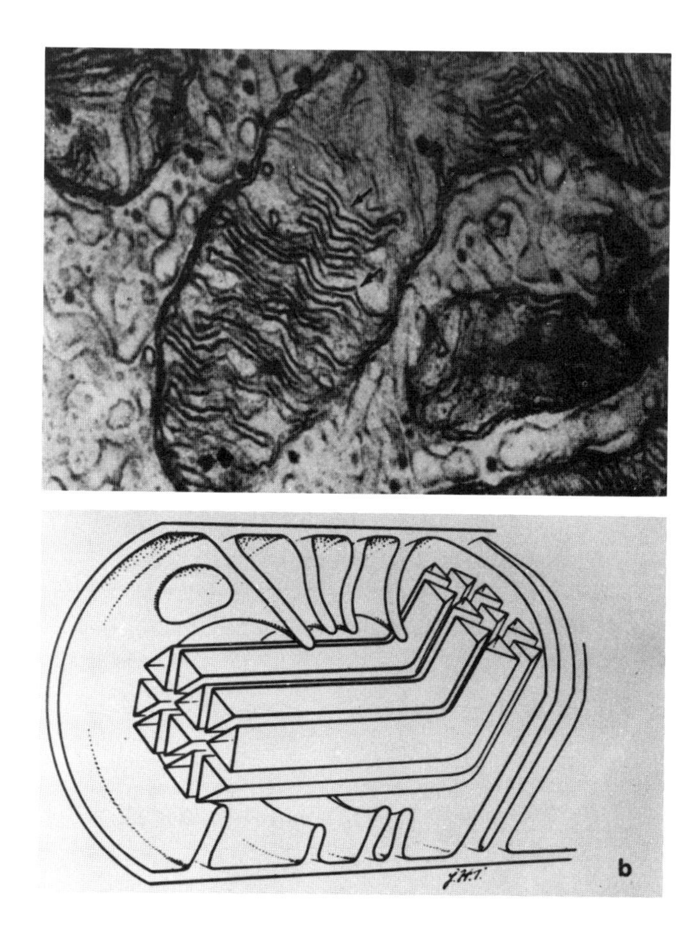

Figure 5. Cristae

Figure and caption from L.T. Threadgold, *The Ultrastructure of the Animal Cell*, Second Edition, p. 327. Oxford: Pergamon Press, 1976. Reprinted with permission

(a) Angular cristae (arrows) in the mitochondria of the chloride cells of *Fundulus*. (x 70,000. Courtesy Dr. C.W. Philpott and the Rockefeller University Press.) (b) Diagram of prismatic cristae from the crycothyroid muscle of the bat. The cristae are hexagonally packed. (Courtesy Dr. D. Fawcett.)

2.3 Models

Some multiple image displays, such as Figures 3–5, include a second kind of diagram to the tracings discussed above. In addition to showing photo-diagram pairings, these displays include diagrammatic "models" which transcend the perspectival limits of the photographs with greater freedom than do the tracings. Unlike tracings, these pictures are not concretely tied to any particular photograph. The endoplasmic reticulum (Figure 3), the ribosome (Figure 4), and crystae (Figure 5) are represented in the models as *generalized* profiles based on a synthesis of particular profiles (Figure 4) or the construction of a paradigmatic 'instance' (Figures 3 and 5).

The "models" shown in Figures 3–5 utilize various representational conventions to produce the illusion of three-dimensionality, and of the exposure of interior detail.[12] These conventions work together to produce an image that is more comprehensive in what it shows, and more 'theoretically' informed than the renderings discussed earlier. Dots or specks become spheres and lines or borders become reticulated membranes (Figure 3). The position of the object is no longer tied to a specific photographic image, nor do its features outline the specific conformations visible in any single photograph. Objects are positioned to reveal multiple sides, and, moreover, to reveal features which are critical for assigning identity and formulating explanation. Rather than further fragmenting the specimen to reveal its details, models reconstruct a holistic entity and seemingly return the viewer to a state of the object before it was analytically disassembled. This view is not, however, the same as the 'original' specimen before it was killed, dissected, stained and otherwise prepared for microscopic viewing. Instead, it provides an imaginary re-assembly of the specimen based upon multiple fragmentary remains (this is graphically illustrated in Figure 4, where the modeled ribosome is displayed as a convergence of partial diagrammatic views).

In addition to merely synthesizing particular representations, certain models (such as Figure 5) are drawn in such a way as to 'expose' internal or underlying 'mechanisms' that serve further to analyze or to explain visible anatomical features. Labels, serial arrangements, and cutaway views display hypothetical processes occurring within the visible structures which cannot concretely be seen in any photograph. The concrete representation of the anatomical entity begins to crystallize not only what can be *seen* of it in various micrographs at a comparable level of magnification, but also what can be claimed about its biochemical structure; a structure which could not possibly be viewed by the same means as the anatomical outlines. Note how the cut-away view of the cristae (Figure 5) seemingly exposes an abstract geometrical arrangement in its inner mechanism. This juxtaposes anatomical context and explanatory geometrics within an integrated visual account; a visual image that nowhere could be seen or photographed in a unified way with currently available techniques other than through the artful assembly of such a diagram.

The model does not necessarily *simplify* the diverse representations, labels, indexes, etc., that it aggregates. It adds theoretical information which cannot be found in any single micrographic representation, and provides a document of phenomena which cannot fully be represented by photographic means. It integrates and assembles the visible, normative, and mathematical products of diverse research projects.

A diagram (and, in its own way, a photograph) is also a model relative to other possible renderings. Modelling is, to an extent, accomplished by a diagram's use of clarified outlines, paradigmatic viewpoints, labels and arrows. Similarly, a micrograph depends on stains (labels), sectional orientation, and photographic framing.

Reasoning and vision are intimately associated from the beginning of the rendering process to its end. It is only by comparing one stage of that process to another that we can distinguish relative degrees of eidetic vs. empirical form. As we trace through the sequence of renderings, we see that the object progressively assumes a generalized, hypothetically guided, didactically useful, and mathematically analyzable form. It becomes progressively less recalcitrant to the textual devices of describing, displaying, comparing, causally accounting, mapping, and measuring.

So far, we have examined features that identify and integrate visual particulars with disciplinary eidos. We will now examine a predominant variant of such identification, that of visualization and mathematical analysis.

3. Mathematization

Illustrations in scientific texts seldom merely *depict* specimens, they integrate the individual and aggregate properties of specimens with mathematical operations. Tables and graphs abound in scientific publications. These visual documents integrate the substantive, mathematical, and literary resources of scientific investigation, and create the impression that the objects or relations they represent are *inherently* mathematical.

For Husserl (1970), mathematization was an historical accomplishment, though he analyzed that accomplishment not as a development in factual history, but as a retrospective implication of "internal history."[13] Husserl's "historical" reflections began with the intuitively taken-for-granted mathematical sciences. From that contemporary starting point he programmatically outlined the problem of how the things encountered in a pre-scientific praxis were measured by, and identified with, the limit forms of geometry:

> First to be singled out from the thing-shapes are surfaces – more or less "smooth," more or less perfect surfaces; edges, more or less rough or fairly "even"; in other words, more or less pure lines, angles, more or less perfect points; then, again, among the lines, for example, straight lines are especially preferred, and among the surfaces the even surfaces, straight lines, and points are preferred, whereas totally or partially curved surfaces are undesirable for many kinds of practical interests. Thus the production of even surfaces and their perfection (polishing) always plays a role in praxis. (Husserl, 1970:376)

The limit forms of geometry, initially used as points or lines of reckoning for guiding constructive action, were cognitively transformed into the basis for a Galilean physics; a physics that posited mathematical order to be the essential underlying *nature* of the empirical world.

Husserl suggested that this historic accomplishment is latent in the structure of the present-day tradition of geometry and mathematical science, although he was not explicit about how this accomplishment is apparent in the ordinary practices of contemporary scientists other than as a set of handed-down presuppositions about natural order.

It is possible to read Husserl's account as a description, not of a once-and-for-all historical movement from proto-science to science, but as an account of what scientists do every time they prepare a specimen for analysis in actual laboratory work. Starting with an initially recalcitrant specimen, scientists work methodically to expose, work with, and perfect the specimen's surface appearances to be congruent with graphic representation and mathematical analysis.[14]

An examination of scientific practices or, as will be presented here, of selected visual displays, can show that the accomplishment of "mathematization" is not buried in the constitutive acts of a "proto-geometrer," nor only in the presuppositions of contemporary scientists. Instead, it is an overt and methodical accomplishment at the ordinary sites of scientific work. To say this is to claim more than that scientific work involves measurement and descriptive instruments which display, map, and explain phenomena in terms of Cartesian coordinates and mathematical equations. It is to point to how instrumental and graphic facilities are implicated in the very organization of what the specimen consists of as a scientific object. The details of laboratory work, and of the visible products of such work, are largely organized around the practical task of constituting and "framing" a phenomenon so that it *can* be measured and mathematically described. The work of constituting a measurable phenomenon is not entirely separate from the work of measurement itself, as we shall see.

Mathematization is embodied in the graph. The graph has become an emblem of science which even popular advertisements exploit. Graphing a phenomenon identifies the thing or relationship with the analytic resources of mathematics. Just as significantly, it places an account of the thing on paper, or prepares the phenomenon with a practical and social universality; not the cognitive universality of mathematics, but the mundane durability, iterability, and invariance of a textual impression.[15]

The analysis of Figures 6–10 will discuss practices which preserve and transform "natural" residues into analyzable points and lines within Cartesian coordinates; lines and coordinates that are then used reflexively to identify 'mathematical' properties of the original specimens. We will be able to see how the 'pure' limit forms of mathematics, such as the point, line, and two dimensional grid, take on the substantive characteristics both of the literary page and of the specimen's residues. What results is a hybrid object that is demonstrably mathematical, natural, and literary. The graph shows a careful attention to the problem of merging these three fields into a unified rendering of a scientific-object; not a 'natural' object, not an object of mathematics, nor simply a literary fiction.[16]

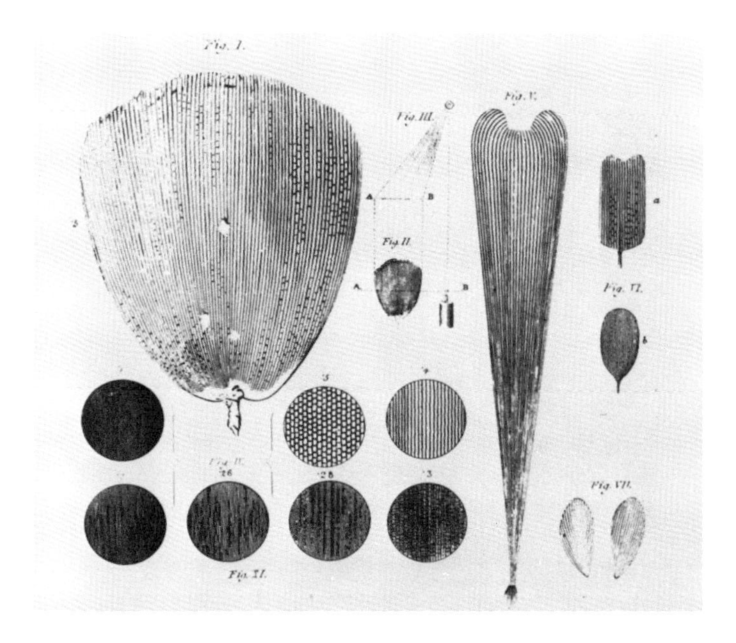

Figure 6. Dr. Goring's test-objects

Figure and caption from G. L'E. Turner, "The microscope as a technical frontier in science," in S. Bradbury and G.L'E Turner (Eds.), *Historical Aspects of Microscopy* (Cambridge: W. Heffer & Sons Ltd., 1967), pp. 175–200. Figure on p. 179. Original in C.R. Goring & Andrew Prithcard, *Micrographia: containing Practical Essays on Reflecting, Solar, Oxy-hydrogen Gas Microscopes; Micrometers; Eye-Pieces, etc., etc.* (London, 1837).

Dr. Goring's test-objects. A scale from the wing of *Morpho menelus* seen through an achromatic objective of nearly 1 inch focal length. The circles show what can be seen when the aperture is increased from 0.1 to 0.5 inch.

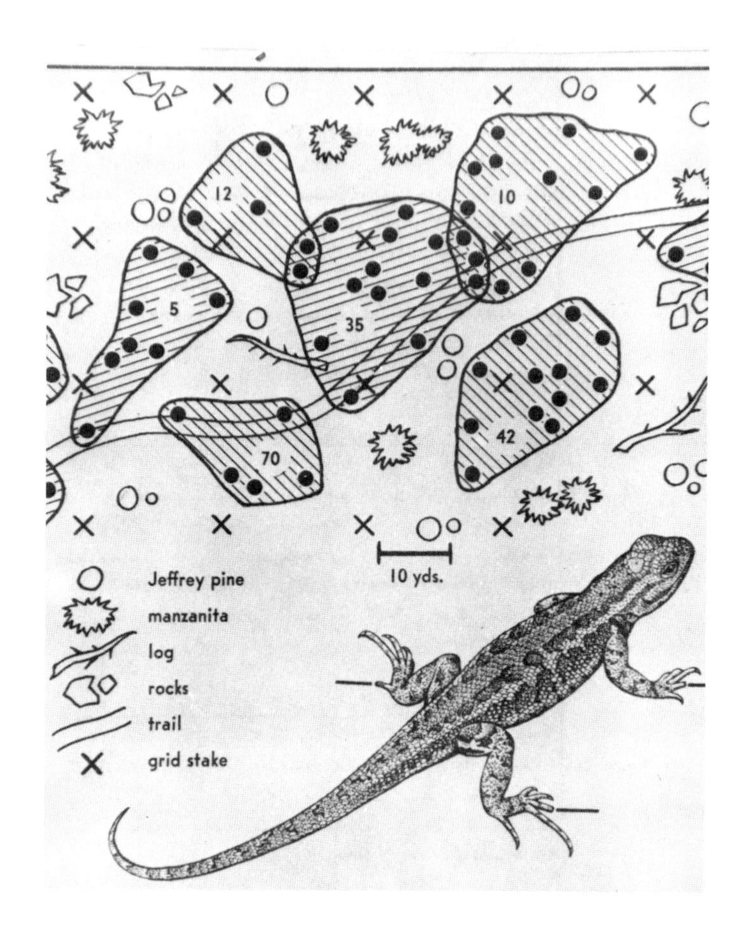

Figure 7. Map of study area showing marked lizards

Figure and caption from Robert C. Stebbins, *A Field Guide to Reptiles and Amphibians*, p. 21. (Houghton Mifflen, Co., 1966). Reprinted with permission

Figure [7] illustrates the kind of results that can be obtained. The home ranges of six Sagebruch Lizards are shown. The lizards were given identification numbers and were permanently marked by removal of two or more toes in combinations (see illus.). Recaptures (black dots) were plotted in reference to numbered stakes (X) arranged in a grid at 20-yard intervals, and were located by pacing the distance to the two nearest stakes. Toes removed from the lizard shown are right fore 4, right hind 3, left hind 5, expressed as RF4-RH3-LH5.

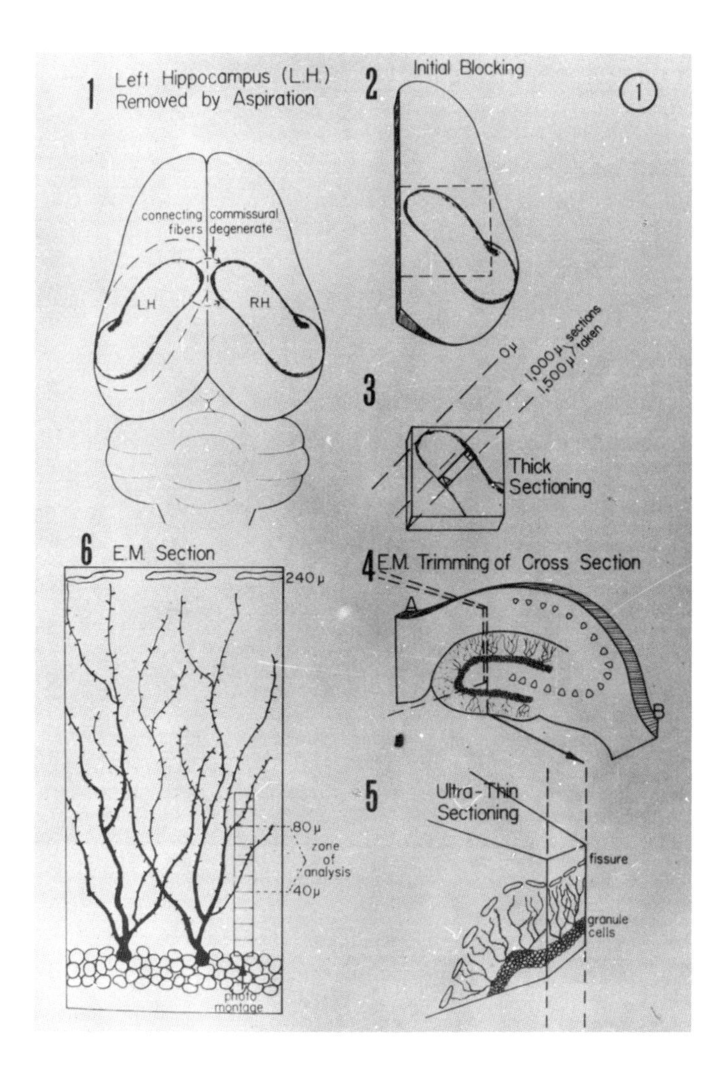

Figure 8. Procedures with hippocampus

Figure and caption from Randall McWilliams and Gary Lynch, "Terminal proliferation and synaptogenesis following partial deafferentation: The reinnervation of the inner molecular layer of the dentate gyrus following removal of its commissural afferents." *Journal of Comparative Neurology* 180 (1978), 581–616. Figure on p. 583. Reprinted with permission

Procedures 1–6: (1) In the experimental animals the left hippocampus is removed completely by aspiration in order to ensure total degeneration of the commissural fibers projecting to the contralateral hippocampus. (2) Following appropriate survival times, the animals are perfused, the brains removed, and the right side blocked. (3) The block is then placed in agar and sectioned on a tissue chopper. Sections 125–150 **u** thick are taken from a region 1,000–1,500 **u** caudal to the rostral tip of the hippocampus. The sections are cut perpendicular to the longitudinal axis of the hippocampus. (4) These sections are placed in capsules containing Epon-Araldite. Following hardening in an oven, the sections are blocked and trimmed for EM sectioning. ... (5) The sections are trimmed so that when placed in the ultramicrotome, they will be cut perpendicular to the granule cell layer and parallel to the dendritic field of these cells. (6) Following routine staining, the ultrathin sections are placed on a non-meshed grid. Photomontages are taken perpendicular to the granule cell layer and the micrographs from the 40–80 **u** zone are analyzed.

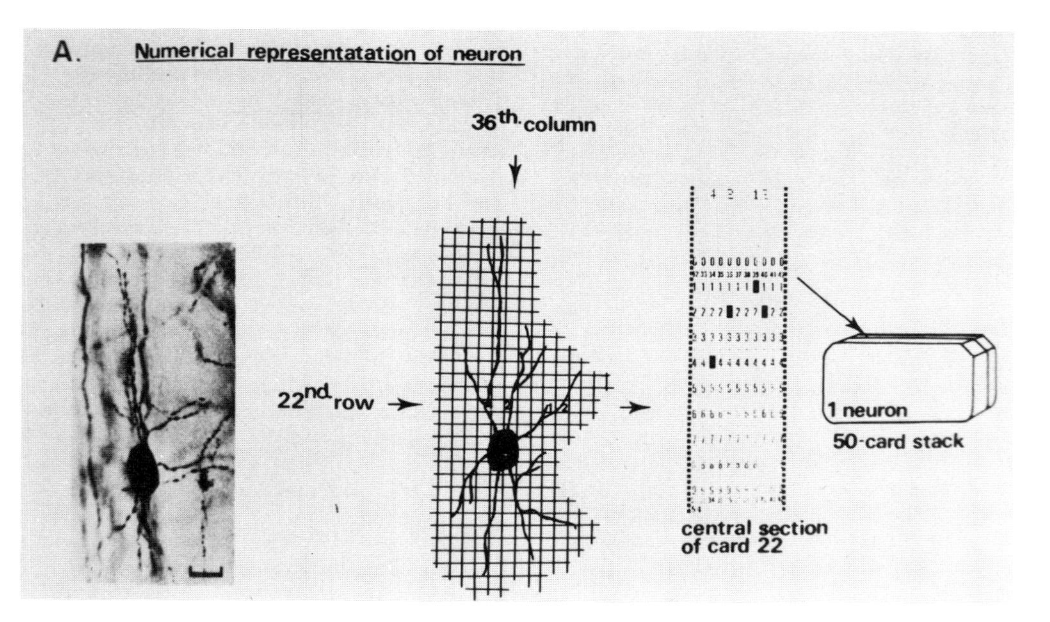

Figure 9. Numerical representation of neuron

Figure and caption from S. Borges and M. Berry, "The effects of dark rearing on the development of the visual cortex of the rat," *Journal of Comparative Neurology* 180 (1978), 277–300. Figure on p. 281. Reprinted with permission

On the left is a typical stellate cell in the visual cortex of a normally reared rat (photographed from a coronal section, 150 um thick of a Golgi-cox impregnation, showing impregnation marker = 10 um sparsely spinous and varicose dendrites which are characteristically radially oriented running parallel with apical dendrites of pyramidal cells. Stellate cells in the visual cortex of dark-reared rats could not be distinguished qualitatively from stellate cells in normally reared rats and thus all cells were numerically represented in preparation for quantitative analysis using scoring system ... and entered into the computer files by punching data in the first 70 columns of 50 IBM cards, each card representing a single row of the grid. A stack of 50 cards thus represents a single neuron. The 3,500 (50 X 70) squares of the grid have a total area of 250 um X 350 um. Each square of the grid had a side length of 5 um.

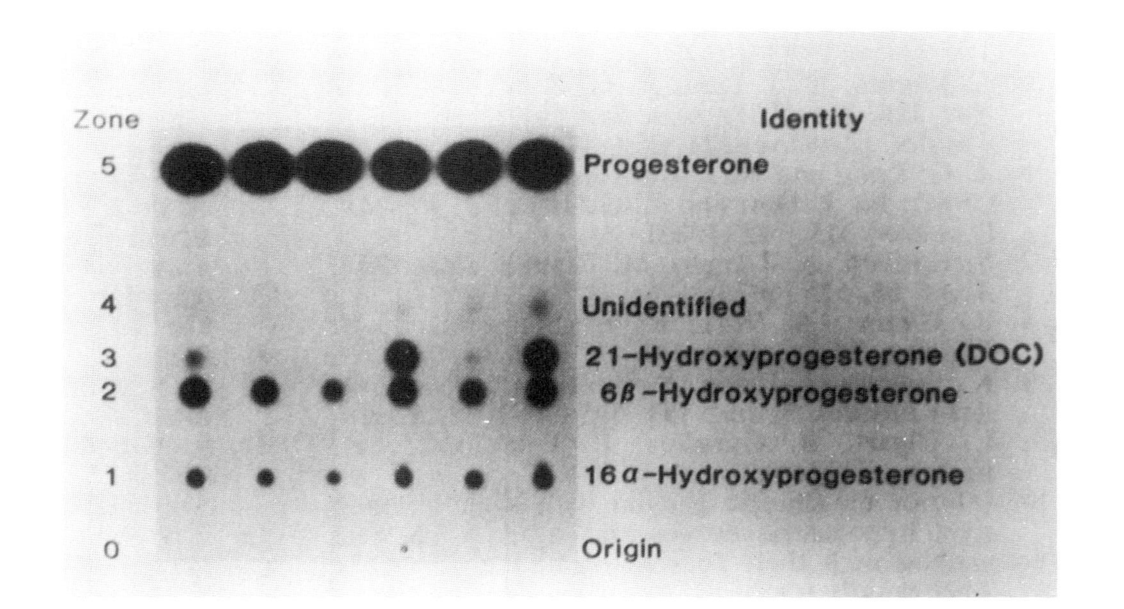

Figure 10. Autoradiographic image

From Herman H. Dieter, Ursula Muller-Eberhard, and Eric F. Johnson, "Rabbit hepatic progesterone 21-hydroxylase exhibits a bimodal distribution of activity," *Science* 217 (20 August, 1982).

Autoradiographic image of the separation of progesterone and its metabolites by thin-layer chromatography. Each lane displays metabolits extracted from reaction mixtures containing equal amounts of liver microsomes prepared from individual New Zealand White rabbits. Three of the principal metabolits are identified by the legend on the right. Assay procedures are described in the text.

3.1 Natural laboratories/Natural graphs

Figures 6–10 are all taken from published works in the life sciences: Figure 6 from a 19th century text of micrographia, Figure 7 from a field guide for studying reptiles and amphibians, Figures 8 and 9 from neurosciences research publications, and Figure 10 from a biochemistry article. Each in its own fashion illustrates how "naturally occurring"[17] fields, objects, or substances are organized into graphic data.

Scientists sometimes speak of "natural laboratories;" unusual circumstances which permit controlled observation, comparison, and experimentation to be performed on fields, objects, or relationships which usually occur in more confounding circumstances. Endless variations on this theme occur in the "descriptive sciences." Astronomers take advantage of eclipses to more precisely measure the shapes and dimensions of eclipsed or eclipsing bodies, and to test models on the astrophysical relationships between such bodies; geologists, paleontologists, and archeologists make methodic use of exposures (such as outcroppings or mineshafts) of sedimentary strata, treating the relative linearity and repeated distributions of strata as temporal scales and indices.[18] Such fortuitous events are not used passively. They are appropriated, and then further analyzed, shaped, and framed to elucidate and upgrade their potential as grounds for comparison, measurement, and testing.

Even within an "unnatural" laboratory, researchers endeavor to select and isolate purportedly "natural" features of their specimens that facilitate controlled observation. For instance, areas of the brain which exhibit a "stratified" anatomy are dissected for analysis instead of regions where the spatial arrangement of cells does not distribute into relatively uniform regions.[19] In other words, there is a preference for specimen materials that exhibit approximate geometricity, not because geometric form by itself is of interest, but because such form provides a convenient basis for specific practical actions. Figure 6, "Dr. Goring's test objects," illustrates an early case in microscopy of a collection of objects which exhibited unusual linearity and regularity of magnified details. The details were not so much of interest in themselves, as they were as naturally occurring grids for elucidating the optical accuracy of the instruments used to examine them.

3.2 Constituting graphic space

In addition to selecting such proto-geometrical objects of study, scientists devise instruments and methods for *upgrading* and *framing* the approximate linearity, uniformity, and regularity of the selected materials. A crude analogy would be a parking lot where builders initially select a relatively flat and rectangular plot of land and then enhance its flatness and rectangularity by introducing a bulldozer into the equation. In Figures 7–10, however, the end product is a graph and not a parking lot. Graphic space is constituted in each case as material fields, or specimen residues are turned into displays of their 'mathematical' properties.

Figure 7 is an illustration from a field guide for identifying Reptiles and Amphibians (admittedly not a bona-fide 'scientific' text, but a popularization which instructs a naturalistic pursuit akin to birdwatching). The illustration provides a clear case of the simultaneous constitution of a mathematical, natural, and literary display.

In this case, a graph is impressed into the very natural terrain with which the observation begins. As the caption to Figure 7 states, an array of stakes is driven into a relatively flat plot of ground to form a grid. The ground thus becomes simultaneously a natural habitat for the lizards and a mathematical field for objectively reckoning the positions and movements of the lizards. Lizards are given numerical identities by *marking* each individual as a unique number in a series. The marks appropriate the naturally occurring 'digits' of the lizards toes, as a unique combination is amputated for each specimen. The lizards are then recaptured over and over again, and during each episode a position is marked on a map with reference to the fixed points of the grid. Over time these positions are diagrammatically summarized into "territories."

On the basis of these transformations, lizard territories take on graphic shape; they are discrete, outlined, and defined by relations of separation and overlap. These spatial properties *literally* come into being on the graphic map reproduced in the figure; a map whose correspondence to a more original terrain depends upon the work of constituting graphic space and numerical markings at that site.

Figure 8 is a diagram from a neurosciences publication. The diagram illustrates a serial process through which a fragment of brain is extracted and sectioned in preparation for electron-microscopic observation (see Lynch 1985a, 1985b, for a more extensive discussion of this case). Note how the sections of tissue are cut as rectangular planes, oriented cross-sectionally and aligned so that the lower edge of the plane approximately parallels an exposed cell layer. It is no accident that the selected tissues appear regularly layered, and that microtomes slicing along fixed planes of operation are utilized to expose and frame the approximate layers. Subsequently, tissues are photographed, and the straight edges of the photographs are turned into scales for marking the plane into a grid (see procedure #6 in Figure 8). Variations in the anatomy of the tissue are then located in sectors at variable distances "above" the cell layer, and the results of the examination are plotted onto a graph representing an aggregate of cases. "Distance above the cell layer" becomes incorporated into distance above the X axis on the graph; "points" on the graph are inscribed by encircling profiles of cellular organelles with color-coded outlines.

Figure 9 provides a schematic account of how a light-microscopic photograph of cells is traced and mapped onto a grid for a study of the visual cortex of the rat. The rows and colums of the grid then provide a matrix for coding the presence or absence of neuronal material in each sector. In this way, anatomical visibility is translated into a numerical code stored on stacks of computer cards for later reference.

3.3 Hybrid 'points' and 'lines'

Complex variations on the basic Cartesian graph occur in science; more than I can treat here or even hope to comprehend. However, one theme which applies to many, if not all, graphs is that of how the commonplace resources of graphic representation come to embody the substantive features of the specimen or relationship under analysis. Commonly used graphic resources consist of an X and Y axis, scales superimposed on each axis, and arrays of points or lines inscribed within the graph's coordinate space. Numerous literary conventions are utilized to identify points, lines, and scales with empirical properties of objects or relationships. Dashed and solid lines endowed with different color and thickness introduce classification and comparison into coherent graphic representation. Similarly, points in graphic space are labeled, colored, and provided with dimension or shape in order to analytically encode that space. Scales are devised variously to provide a map of a concrete sector of space, or to represent a relationship between quantitative or qualitative variables.

As discussed above, efforts are made to shape specimen materials so that their visible characteristics become congruent with graphic lines, spaces, and dimensions. Scales and axes are superimposed upon the rectangular frame of a microscopic photograph so that the inspectable details of the photograph can be marked and counted as a distribution of points within different sectors of a grid (Lynch, 1985a). These graphic points and locales are traced back to the material residues of the specimen.

Figure 10, a photographic reproduction of an "autoradiograph," clearly illustrates the practical integration of 'material' and 'mathematical' order within the representational conventions of a graph. Specimen residues (in this case, molecular extracts labeled with radioactive markers) are instrumentally processed to resemble points, though they do not become "pure" points in the sense of representing dimensionless locales. The dimension that distinguishes the empirical spots from points is highly significant for what the graph communicates. We see different dots discriminated by size lining up as though ordered in a table. Although they act as points within a spatial grid, they simultaneously bear categorical identities and are 'listed' as though they were tabular entries.

Although the spots are not dimensionless points, neither are they simply pictures of molecules. The identities of the spots as specific kinds of molecules are supplied by their placement in the grid as well as by their relative dimensions. The taxonomic table is integrated with the instrument of observation, so that the instrument not only uses the table to record the identity of the specimen, it arranges the observable qualities of the object to be essentially graphic in character. The specimen extracts are filtered through a cleverly devised instrument constructed to discriminate visible molecular products into lanes and rows. Each lane identifies an individual specimen, and each row identifies a specific "molecule" analytically isolated from the specimens. Molecular identity is visible as nothing more than the two dimensional size of the spots and their graphic distribution. Although more substantial than points, the "molecules" are also highly abstract, taking the form of opaque spots identified with graphic codes.

Variations on the commonplace organization of the graph are endless: Are points plotted on paper 'by hand' or are they meant to identify substantive properties of specimen residues (such as in Figure 10)? Do lines and scales originate from the designs of instruments (Figure 10), or do they claim a material relation to anatomical strata (Figures 6 and 8)? Hybrid graphic features can be viewed as relatively abstract or concrete; as originating through literary representation or 'hands off' instrumental transfers from 'world' to paper (such as with a seismograph). The rhetorical impact of a graph, as an empirical document, representation of a theory, or illustration of an argument can depend strongly on the apparent origin of its features. In all cases, such hybrid features are neither wholly mathematical nor empirical. In either case, they are literary objects, since they inevitably end up on paper and are defined by conventions of graphic representation.

The mathematical/natural/literary unity of particular displays *can* be taken apart. Data appearing as graphic inscriptions can be treated as detachable signs, independent of the objects to which they refer. This is often advantageous, as long as their 'natural reference' can be presumed, since extrapolations *about the object* can be made by operating on the graphic features as though they consisted entirely of mathematical entities (i.e., numbers, lines with a measurable slope, a distribution of points). Under some circumstances, however, the presumed reference of graphic features to "natural" order is called into question, and graphic entities come to be seen as artifacts or possible artifacts (Lynch, 1985b). Although artifacts are fairly common in day to day laboratory research, their existence does not markedly inhibit the programmatic treatment of features of graphic visibility *as* the sensual properties of scientific objects. There is little point in doing otherwise since the "original" objects of microscopic research, for instance, are always hidden until they are made observable through the artifices of staining, sectioning, magnification, and the devices of visual and graphic representation discussed above (cf., Hacking, 1983: Ch. 10).

4. Conclusion

This paper discussed two themes: selection and mathematization. The first theme concerns methods for selecting and retaining visual/evidentiary elements of a specimen in subsequent renderings of it. It was argued that scientific representation is more than a matter of reducing information to manageable dimensions. Representation includes methods for *adding* visual features which clarify, complete, extend, and identify conformations latent in the incomplete state of the original specimen. Instead of reducing what is visibly available in the original, a sequence of reproductions progressively modifies the object's visibility in the direction of generic pedagogy and abstract theorizing. We see this latter set of operations especially clearly in "models" which depict 3-D, color-coded, and labeled expansions of black and white, unadorned cross-sections. In addition to adding the illusion of a third spatial dimension, models enable certain theoretic relations to be represented as though they were "in" the depicted objects. Models use visual conventions to denote change, sequence, activity, and relationships of various kinds. The object becomes more *vivid*; we can picture it as though it were "naturally" present for our inspection.

Mathematization includes practices for assembling graphic displays. Specimen materials are "shaped" in terms of the geometric parameters of the graph, so that mathematical analysis and natural phenomena do not so much *correspond* as do they *merge* indistinguishably on the ground of the literary representation.

The two themes of selection and mathematization are intimately related. Preparation of an object for mathematical operations includes and builds upon many of the practices discussed under selection/simplification, such as clearly marking outlines to distinguish one case from another, and providing arrays of specimens with uniform categorical identity. These methods produce objects that are coded and implicitly aggregated. Arithmetic and graphic representational operations can then be performed on the basis of the enhanced identities and differences. Constructing a "good" picture of a laboratory specimen's residues is therefore prerequisite for mathematically analyzing those residues.

The dependency of mathematical data upon such preparatory practices can, in particular cases, be exposed as a source of error. However, this does not necessarily imply that they are 'suspect' in general. The point is that the practices are necessary for constituting data, whether or not they are seen to be a source of error, and that it is only when they are taken for granted that the attribution of mathematical order to "nature" can succeed.

Notes

1. In addition to the papers in this special issue, sociological and historical studies which deal specifically with visual images, their construction, and their interpretation in science include: Alpers (1983), Bastide (1985), Edgerton (1976), Ferguson (1977), Gooding (1986), Gilbert and Mulkay (1984: Ch. 7), Ivins (1973), Jacoby (1985), Latour (1986), Lynch (1985a), Morrison (1988), Rudwick (1976), Shapin (1984), and Tilling (1975). See Latour and De Noblet (1985) for an edited collection of studies on "visualization" in science.
2. Latour (1986), drawing on recent developments in the sociology of language, argues for a revision of the cognitive sciences which substitutes the external domain of the literary text for the internal realm of mind as a locus of sense and reasoning.
3. For a discussion of how the scientific object can be viewed as a "cultural object," when account is taken of the local practices of discovering and transforming the object in scientific research, see Garfinkel et al. (1981).
4. Arguments on why scientific texts cannot be treated as descriptions of laboratory practices are presented in O'Neill (1981), and Knorr-Cetina (1981).
5. Morrison (1988) demonstrates how visual displays are incorporated into a scientific text's reading, enabling the reader to see what the text is saying.
6. Gurwitsch (1964) makes reference to James (1905: Ch. 9, Section 5). Also see, Schutz (1962:5).
7. The theme of selection or simplification is discussed in reference to scientific practices in Star (1983). The theme is used, although less centrally, in Latour and Woolgar (1979), and in Callon et al. (1985).

8. That different instrumental displays represent a "same thing" is, of course, a highly problematic claim. It is problematic not only in an epistemological sense, but also in a practical sense. For laboratory practitioners, the 'objective' reference of different renderings is a standing technical problem. When I mention different renderings of "a same thing" in the passage above, I mean nothing more than "sameness for all practical purposes." Identity in such a case is contingent and defeasable.
9. Garfinkel (forthcoming) speaks of "rendering" practices in a number of sociological contexts, including science. "Rendering" are transformed when, for instance, speech is written down or embodied actions are tape recorded. The sequential relation described here by the term is vaguely reminiscent of the theme of "conditional relevance" used to describe the sequential ties between conversational utterances. See Sacks et al. (1974).
10. For discussions of spectacular perceptual/diagrammatic artifacts such as Hartsoeker's homunculi pictured within sperm cells, and globulist theories of cellular structure see Bradbury (1967), Ritterbush (1972), and Turner (1967).
11. By "eidetic image" is meant, not a "mental picture," but an image that synthesizes the eidos of a field or discipline. The term is adapted from Husserl's philosophy, where it is used to refer to the transcendental 'essence' of an object-in-experience. Here, it is stripped of its transcendental overtones, and refers more concretely to the generalized or idealized version of an object portrayed in a visual document.
12. Representational conventions used in scientific illustrations are discussed in Alpers (1983), Edgerton (1976), Ivins (1973), and Rudwick (1976).
13. See two sections of Husserl (1970): "Galileo's mathematization of nature," (pp. 23–59), and "The origin of geometry," (pp. 353–378).
14. The idea of a "local history" of scientific work is given original development in Harold Garfinkel's work. See Garfinkel et al. (1981).
15. Latour (1986), and Callon et al. (1985) implicate the scientific text in the development of scientific communities. Because a text can be circulated throughout a dispersed literary community, and because what a text says is "immutable" (or at least iterable in Derrida's sense), Latour argues that scientists are able to constitute a sense of scientific fact and method, as well as scientific community itself, as a transsituational phenomenon. Also see Goody (1977).
16. This presents an interesting problem: why isn't the graph a fiction? O'Neill (1981) argues that scientific writing cannot formally be distinguished from fictional writing, and Gilbert and Mulkay (1984: Ch. 7) speak of scientific illustrations as "working conceptual hallucinations." Graphs in, for instance, magazine advertisements utilize the form of Cartesian coordinates to make the case for products like low-tar cigarettes, and even perfumes. It is not enough to say that the advertisement is a fictional creation, while the scientific graph represents some-

thing in 'nature.' However, analyses of scientific practices indicate that an immense amount of care is taken in laboratories to 'package' material residues of a specimen into graphic form. Such work is, of course, susceptible to fraudulent uses, and there is no ultimate assurance that the naturalistic claims embedded in a graphic display are non-illusory. But, we can unpack how these claims attempt to encompass natural residues whether or not they successfully do so.

17. "Naturally occurring," is used in the phenomenological sense. It does not imply an object unaffected by human perception and action, but an object that has not yet undergone the specific transformations produced by scientific practices.
18. Rudwick (1976) discusses the significance of exposed strata in the history of geology. An example of contemporary archaeologists' use of strata is the following (Kirk and Saugherty, 1978):

> As part of the permanent documentation of archaeological deposits actual columns can be pulled from excavation walls with all layers intact. These permit convenient reference to the precise sequence and nature of deposits even long after field work is completed. The technque is simple. Let resin soak into a vertical section of wall, usually a place where strata are particularly clear and have special significance. When the resin has hardened, cut the sides of the column free, let more resin soak in, and place a board against the face of the section. Gingerly cut the columns free from the wall and tie it to the board. Wrap with burlap or plastic sheeting and take the whole to the laboratory for additional resin treatment and study. Such columns form a permanent "library" of stratigraphy.

19. The practical advantages of the hippocampus (an anatomically 'stratified' area of the brain) are discussed in G. Lynch et al. (1975).

References

Alpers, S. (1983). *The art of describing*. Chicago: University of Chicago Press.

Bastide, F. (1985). Iconographie des textes scientifiques: Principes d'analyse. *Culture Technique* (14):132–151.

Bradbury, S. (1967). *The microscope, past and present.* Oxford: Pergamon Press.

Callon, M., Law, J., and Rip, A. (1985). *Qualitative scientometrics: Studies in the dynamic of science.* London: Macmillan.

Edgerton, S. (1976). *The Renaissance discovery of linear perspective.* New York: Harper and Row.

Ferguson, E. (1977). The mind's eye: Non-verbal thought in technology. *Science* 197:827ff.

Garfinkel, H. (forthcoming). *A manual for the study of naturally organized ordinary activities*. London: Routledge and Kegan Paul.

Garfinkel, H., Lynch, M., and Livingston, E. (1981). The work of a discovering science construed with materials from the optically discovered pulsar. *Philosophy of the Social Sciences* 11(2):131–158.

Gilbert, G.N., and Mulkay, M. (1984). *Opening Pandora's box: A sociological analysis of scientists' discourse.* Cambridge: Cambridge University Press.

Gooding, D. (1986). How do scientists reach agreement about novel observations? *Studies in History and Philosophy of Science* 17.

Goody, J. (1977). *The Domestication of the savage mind.* Cambridge: Cambridge University Press.

Gurwitsch, A. (1964). *The field of consciousness.* Duquesne: Duquesne University Press.

Hacking, I. (1983). *Representing and intervening: Introductory topics in the philosophy of natural science.* Cambridge: Cambridge University Press.

Husserl, E. (1970). *The crisis of European sciences and transcendental phenomenology*. Evanston: Northwestern University Press.

Ivins, W. (1973). *On the rationalisation of sight.* New York: Plenum Press.

Jacoby, D. (1985). Vulgarisation et illustration dans les sciences de la vie. *Culture Technique* (14):152–163.

James, W. (1905). *The principles of psychology*. New York.

Kirk, R., and Daugherty, R.D. (1978). *Exploring Washington Archaeology.* Seattle and London.

Knorr-Cetina, K. (1981). *The manufacture of knowledge: An essay in the constructivist and contextual nature of science.* Oxford: Pergamon Press.

Latour, B. (1986). Visualisation and cognition: Thinking with eyes and hands. *Knowledge and Society: Studies in the Sociology of Culture Past and Present* 6:1–40.

Latour, B., and De Noblet, J., Eds. (1985). *Les "vues" de l'esprit*. Special Issue of *Culture Technique* (14) (June).

Latour, B., and Woolgar, S. (1979). *Laboratory life: The social construction of scientific facts*. London: Sage.

Lynch, G., Rose, G., Gall, C., and Cotman, C.W. (1975). The response of the dentate gyrus to partial deafferentation. *Golgi centennial symposium proceedings*. New York.

Lynch, M. (1985a). Discipline and the material form of image: An analysis of scientific visibility. *Social Studies of Science* 15:37–66.

Lynch, M. (1985b). *Art and artifact in laboratory science: A study of shop work and shop talk in a research laboratory*. London: Routledge and Kegan Paul.

Morrison, K. (1988). Some researchable recurrences in science and situated science inquiry. In D. Helm et al. (Eds.), *The interactional order.* New York: Irvington.

O'Neill, J. (1981). The literary production of natural and social science inquiry. *Canadian Journal of Sociology* 6:105–120.

Ritterbush, P. (1972). Aesthetics and objectivity in the study of form in the life sciences. In G.S. Rousseau (Ed.), *Organic form: The life of an idea.* London: Routledge and Kegan Paul.

Rudwick, M. (1976). The emergence of a visual language for geological science 1760–1840. *History of Science* 14:148–195.

Sacks, H., Schegloff, E., and Jefferson, G. (1974). A simplest systematics for the organization of turn-taking in conversation, *Language* 50(4): 696–735.

Schutz, A. (1962). *Collected papers I: The problem of social reality.* The Hague: Martinus Nijhoff.

Shapin, S. (1984). Pump and circumstance: Robert Boyle's literary technology. *Social Studies of Science* 14:481–521.

Star, S.L. (1983). Simplification in scientific work: An example from the neurosciences. *Social Studies of Science* 13:205–228.

Tilling, L. (1975). Early experimental graphs. *The British Journal for the History of Science* 8(30):193–213.

Turner, G. L'E. (1967). The miscroscope as a technical frontier in science. In G. L'E. Turner (Ed.), *Historical aspects of microscopy.* Cambridge: W. Heffer and Sons.

The iconography of scientific texts: principles of analysis

FRANÇOISE BASTIDE (translated by Greg Myers)

1. Introduction

To introduce my argument, I will take as an example a recent discussion in *Nature* concerning an accusation of fraud that was brought against an article thirteen years after the original publication.[1] The article under attack describes the crystallization of a transfer RNA for valine extracted from yeast. It contained an illustration of the x-ray diffraction pattern reproduced in figure 1. The accusers used this photograph to identify the crystal as that of human carbonic anhydrase. Since that crystal had already been obtained and researched at the time of the article, the "discovery" would have been pure invention. The fraud lay in making a "deliberately deceptive representation"; in trying to pass off a photograph of a known molecule as a photograph of one that had been newly crystallized.

The advantage of studying a case in which fraud is suspected is that both the comments of the accusers and the defense offered by the accused display clearly what it is one normally expects of an illustration, by detailing point by point what conditions are not fulfilled in that case. In contrast, the original text,[2] like any other scientific text, contents itself with saying, "figure *x* illustrates. . . ." The textual analyst can only accept this evidence. The article by Paradies and Sjöquist attempts, in effect, to visualize the three-dimensional structure of a molecule by using X-ray diffraction methods that permit one to resolve the positions in space of the consitutuent atoms. The attempt is not entirely successful; the crystal they obtained dissolved during the exposure to X-rays. However, the preliminary results confirmed that the molecule had an ordered, rigid structure that would allow calculations when the

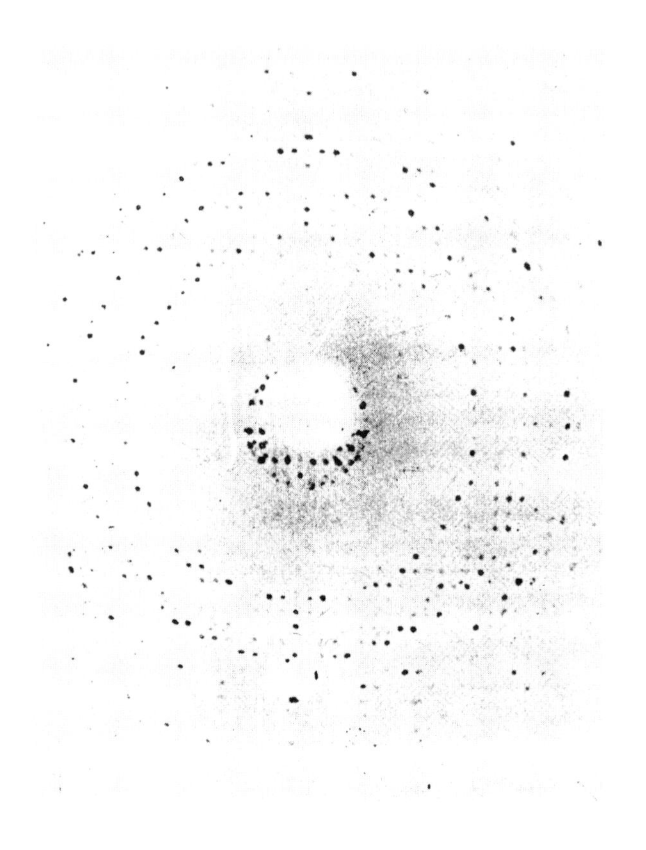

Figure 1

crystallization conditions were exactly right. The article, if it was cited before the controversy that drew my attention to it, was probably reduced to this message. Michel Callon et al. term such a technique a *translation technique*.[3] For convenience, one can conceptualize the article as containing two distinct layers of text; the description of the experimental technique that enabled the results to be obtained, and the general framework of the problem of the structure of transfer RNA. As long as it is not contested the two layers of the article form a whole that I will call a technique of visualization. I would like to stress one of the very general characteristics of the article; it is a matter of using experiment to *make one see* what is invisible. The whole article is a technique of visualization for the public, for the scientific community concerned; the experimental conditions concern a technique of visualization more limited to the researcher's own usage. It is the latter technique that furnishes the material for the published illustration, so it is essentially this second technique that we will discuss. A look at Figure 1 shows that there is a great distance between, on the one hand, spots disposed in concentric circles in the photo, and on the other hand, what one could call the "object," the tRNA molecule that we will see again, drawn artistically, at the end of the present study (Figure 19).

What one sees in Figure 1 is in fact the *trace* of the presence of the molecule, obtained by a technique that is exactly analogous to the study of beds of clay which have preserved the steps of dinosaurs going about their business. In the case of the long-dead dinosaurs, it is the gap between their time and ours that renders them "invisible." By contrast, the tRNA is rendered invisible by its minuteness and by the difficulty of preserving the organization of the molecules in space while preparing them for electron microscopy. The crystal, which, if it existed, would measure $0.2 \times 0.15 \times 0.2$ mm, was visible to the eye, but the disposition of atoms remained invisible. However, as a crystal represents an ordered array of very numerous molecules, it behaves in certain respects like a single molecule enlarged. Nevertheless, it is not the molecule that becomes observable but (as in the case of the dinosaurs) its traces: the deviations, fixed by photography, that the crystal causes in the parallel and rectilinear trajectories of X-rays crossing it. (In addition, the beam was being moved through an arc, but this supplementary detail adds nothing to my point.) The technique makes visible not the object itself, but the result of its action, which I have called, following Bruno Latour, its trace.[4] The crystal arranges the light that crosses it in a new disposition in space, a disposition registered by photography. The result presupposes the action, and the action presupposes a capacity, which is here the capacity to structure light, "in the image" of its own structure. When the chain of presuppositions is not in question, the structure of points in the image makes the ordered structure of the transfer RNA of valine visible.

One could symbolize the technique of visualization in the following schema:

E ▶ O ▶ Op S ▶ P ▶ R

O (for Object) is the invisible structure transformed by the Operating Subject (Op S) into a visible structure (the Product of P); the Emitter (E) can be considered as "nature" (in this case the yeast) and the Receiver (R) is the researcher and, through him or her, the scientific community.

The questioning of the identity of the crystal that has been photographed destroys the unity of this technique, so that it splinters into three experimental components: purification, crystallization, and X-ray technique. The process of writing is thus separated from that of experimentation. Furthermore, in order to understand some elements of the controversy, the simple schema of each one of the now autonomous strategies must be completed by bifurcations that mark the places where an error could be introduced by a substitution. In Figure 2 I present the phenomenon in summarized form: I have adopted a vertical representation, and, in order to highlight the bifurcations, have simplified the writing of the elementary schemas by putting the object and its emitter (or receiver) on the same line. On the left, I have put as a reminder the general schema within which Paradies would certainly have wanted his article to stay.

The method of purification is not a matter of controversy. On the contrary, the critics are careful to exculpate the coauthor of the article they attack, whose role in the project was the purification and who did not know about the crystallography, so that he is not implicated in the fraud.

The conditions of crystallization, on the other hand, are commented on in the editorial by Peter Newmarck, by way of an anecdote. Paradies, visiting a laboratory where attempts to reproduce his protocol of crystallization had not been successful, arrived without his own crystals, claiming they had been lost on the ferry from Germany to Sweden and, according to Clark, failed to produce any more crystals in his month or so at Cambridge. Shortly thereafter Clark succeeded in getting crystals by a modification of the Paradies protocol. Clark published his results in *Nature* (219, 1968, p.

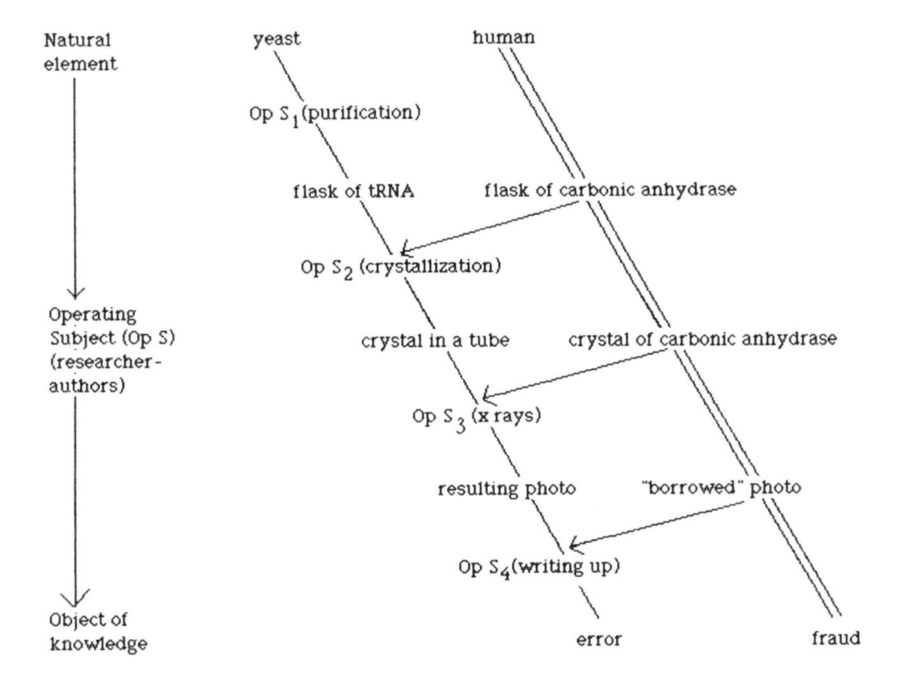

Figure 2

1222), and generously, it might be thought, made Paradies a co-author on the paper for having stimulated the protocol that was eventually successful. This account was intended evidently to throw doubt on the existence of step 2 by calling into question the product, and thus, by implication, the competence of the corresponding Operating Subject, a move which breaks the chain of presuppositions.

The accusers, for their part, suggest (generously, one would say, since error is more pardonable than fraud) the idea that the sample of carbonic anhydrase "had inadvertently been mistaken for one of tRNA." This supposition has the effect of breaking the chain in another way, by introducing a bifurcation at the beginning of step 2.

The identity of the product becomes uncertain. "However, such a mistake could not be the fault since the reported conditions for crystallization differ markedly" from those needed for the carbonic anhydrase.

On the other hand, they do not suggest that Paradies could have confused two crystals when he tested them (which would introduce a bifurcation at the beginning of step 3); as before, the different conditions of the environment of crystallization render that improbable. I have however represented this alternative on my table.

They suggest that the published photograph could have been substituted for the "true" one, by mistake, in the course of step 4, the editing of the article. They think, however, that the author would have noticed it, at least in reading the proofs, and would have rectified it if the substitution had not been intentional. With this suggestion, they make one think that step 3, as well, did not exist, since they have not even envisaged a possible error at this level. However, they indicate that after having identified the molecule, they looked to see if the published photograph was in their archives.

It is surprising, in fact, that the dispute makes a detour by way of the figure instead of the critics simply announcing that the values of two parameters (out of three) calculated by Paradies are not in agreement with the values found later. One can suppose that the values given have been confirmed—Paradies repeats them in his response. The lesson that one can draw from this story is that the precondition for a "successful" fraud (one that is not suspected immediately and is published) is that the invented part stays within the limits of the system's own logic. This takes nearly as much "intuition" as a discovery that would be confirmed; indeed, the ideal is to commit an "appropriate" fraud [*de frauder "juste"*]. But we are not concerned here with fraud as a heuristic. What interests us is the place of the photographic document. After thirteen years, this figure plays a very different role: in the initial article by Paradies and Sjöquist, it is presented as a preliminary result that nevertheless manifests a certain degree of order in the crystal. The plate is however inappropriate for the calculation of the set of parameters defining the structure of the molecule in space, because the caption specifies that it was taken from a somewhat acrobatic shot done under nonstandard conditions. In fact, Paradies indicates that the crystal he has obtained is very labile and that it decomposes under exposure to X-rays (a subsequent figure shows the crystal as missing entirely!). The figure is, then, only there to show the potential reader that the author is well on his way to finding an adequate environment for crystallization. On the other hand, for the later critics the meaning of the photograph changes; they manage to make some calculations on it that the author said were impossible in order to identify the crystal that has produced it. It is in fact by calculating the parameters, taking a standard experiment as the point of departure, and without taking notice of the angles and distances given in the caption to the figure, that the molecule was recognized. What brought the critics to doubt these figures was the symmetry of the image, which one would not expect to find under the conditions given for the shot. On the other hand, Paradies maintains in his response that the calculations made by his critics are groundless.

There is therefore a double movement from the figure to the text and from the text to the figure: the figure, showing the crystal the production of which is explained in the text, validates the experimental protocol; the text, explaining how to read the figure, gives it its meaning. These are the two relations that the two accusers set out to destroy by undermining confidence in the middle step of the experimental technique, which constitutes the sole innovation of the article. The crystal is a different one, and the conditions of crystallization described are only "by chance" applicable to transfer RNA. This unhappy example notwithstanding, we will conclude that in the double movement from the illustrations to the text and from the text to the illustrations, the photograph functions as a *guarantee,* and assures the linking of the various steps in the technique of visualization so that it appears linear and consistent. It also functions as a memory, to the dismay of suspects of fraud or of error.

2. The reading of the image: the channeling of meaning

2.1 Comparison

One could conclude from this story that the photographic document speaks "of itself" (much later, one must recognize), since it is on the figure that the suspicion of fraud settles. However, what can one say about Figure 3? The bed of a river, or a shot of a wall decorated with pebbles? Without the caption that gives the scale (two micrometers for the length of the white bar at lower right) who would imagine that it is of microscopic phosphorus-rich calcium granules incorporated into the muscles of a sea worm?[5] The following figure illustrates another aspect of the decontextualization of images: what is one supposed to see in Figure 4? Simply that the object exists? Is it necessary, rather, to look at the symmetry (or at its lack of perfection?) Is it necessary to take note of the little white spots speckling the surface? A discrete, almost invisible arrow points out the existence of a split in one of the spokes of the upper wheel. The issue is the description of how this structure grew.[6]

These two illustrations present a common point for the analyst of images; both show "something" in contrast with the background. However, in the case of the "pebbles," the "background" (the muscle) is barely noticeable as some horizontal lines; it takes up a very small portion of the surface of the photograph and has no interest except in providing interstices between the granules. On the other hand, in the case of the presentation of the split in one of the spokes of the spine of a sponge presented in Figure 4, two systems of op-

Figure 3

Figure 4

position come into play one after the other: first, as in the other figure but more sharply, the opposition of the spiny object against an undifferentiated background; then the opposition of two objects, the split spoke "against" all the other spokes that are not split. Only the second opposition is pertinent in the article; without the caption, one would not know what to look for. One can say of this photograph either that it carries too much information (the form of the spine on the background, the white marks that speckle its surface, the split spoke) so that the uninitiated reader does not know how to choose the "right" detail (that which matters to the author), or that it contains only very banal information that one can state as: "Here is what the spike of a siliceous sponge looks like."

I must however admit that I have played a trick in presenting these two images in isolation: each one makes part of a series in the article in which it appears. Figure 3 is part of a montage (Figure 5) that includes three other electron microscope photographs before the one I have chosen: the first shows a transverse section of the

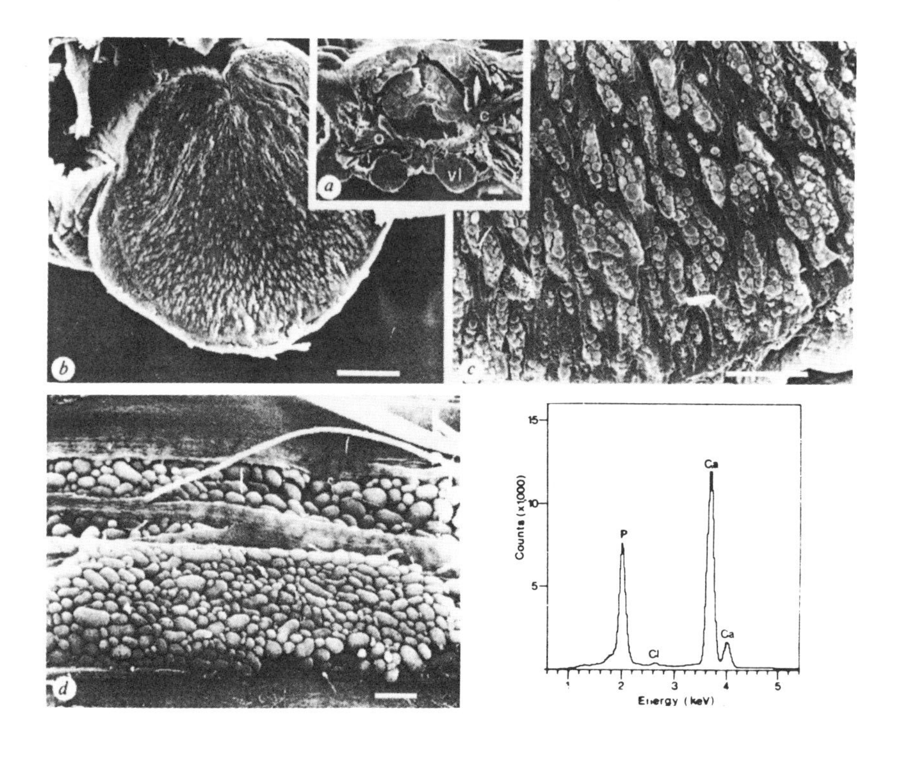

Figure 5

body of the entire worm, the second shows a particular muscle, at a higher magnification, and the third, at an even higher magnification, shows the muscle fibres displaying some inclusions. The fourth, which I isolated at first, shows one of these inclusions on a larger scale. The constitution of the series, then, aims to simulate a progressive focusing of attention on the object examined by the article. Furthermore, the photo presented in Figure 3 is followed with a fifth photo, very different from the those preceding it, the record of an X-ray analysis "saying" that the granules are made of calcium and phosphorus. But I will return later to the analysis of such images. Figure 4 is drawn from a plate (Figure 5) showing two series of spines, the first taken from "control" sponges, above (from which I took Figure 3), the others, below, more or less malformed, taken, the legend explains, from sponges grown in an environment containing germanium. The series, then, aims to show the perturbing effect of germanium on the growth of spines which are made of deposits of the salt of a related metal, silicon.

In the three cases that we have considered, one can remark that the isolated figure takes on a meaning only in an external system of comparison; in fact, the internal comparison of the object with its background was scarcely informative. This external comparison is explicit when Figures 3 and 4 are included in the series of Figures 5 and 6 respectively. But such a comparison is also available in the case of Figure 1 in a more implicit form. On the basis of such documents presented in this figure, the parameters characterizing the crystalline forms of various organic molecules have been calculated and entered in a data base. In the process of searching such a data base, the identity of the molecule represented in the incriminating photo was discovered.

2.2 The force of habit

The reading of Figures 5 and 6 is based on putting them each in a series: by comparing the photos the "significant" differences, the differences that give the key to the reading, become evident. In contrast, Figure 1, which calls on an implicit comparison, was readable only by those accustomed to the technique, the nonspecialist being obliged to trust the text to show what was at stake. A comparison between images, either presented or supposedly memorized, is not always enough to give the image meaning. Often it is also necessary to have a key—or rather, a well-established habit of thought—to be able to interpret a photograph. This point is illustrated in Figure 7, leaving aside for the moment the "meaning" of this illustration in the article from which it is taken. This figure shows two photographs side by side: the photo at left shows a sort of pit (or trench) at top left, and a series of low walls enclosing irregular surfaces. In the photo at right, one can distinguish a network of crevices, with an isolated peak at lower right. The difference between these photos is an optical illusion: the impression of relief is due to the

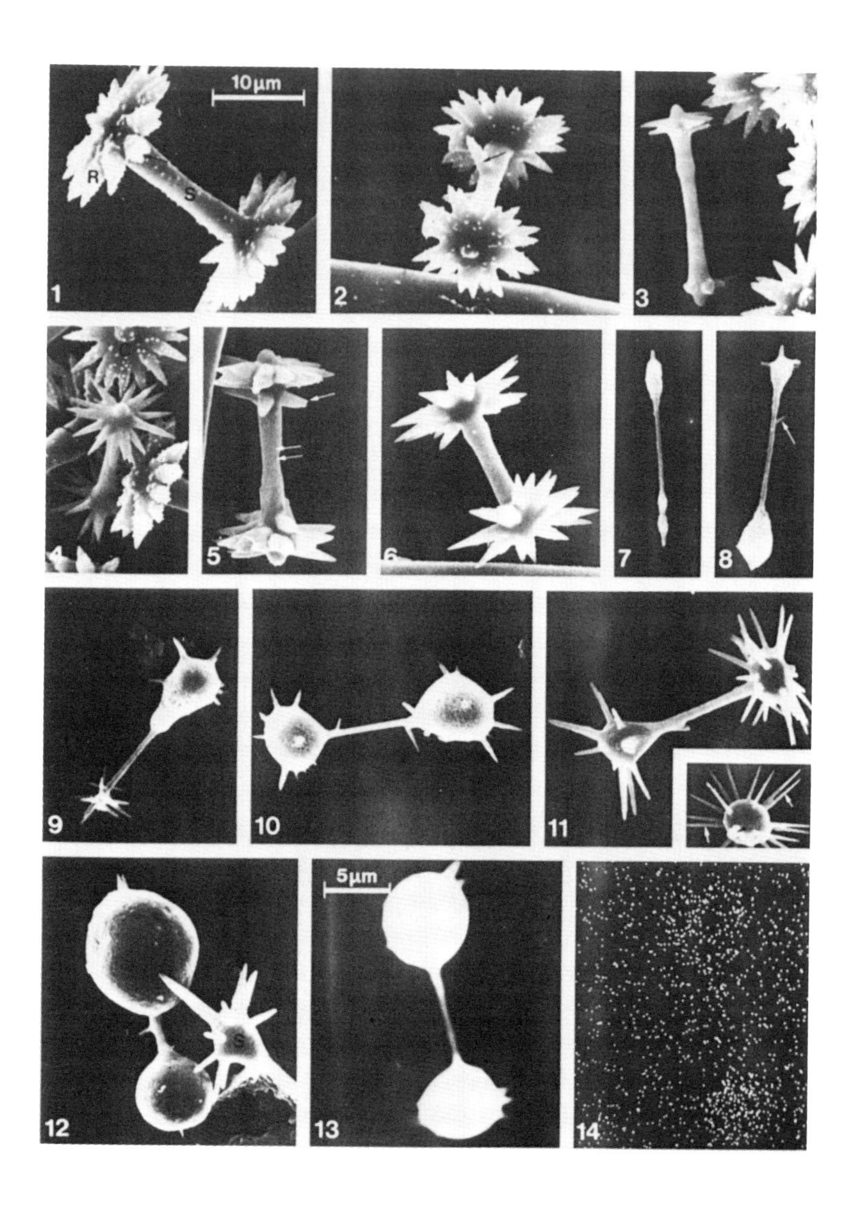

Figure 6

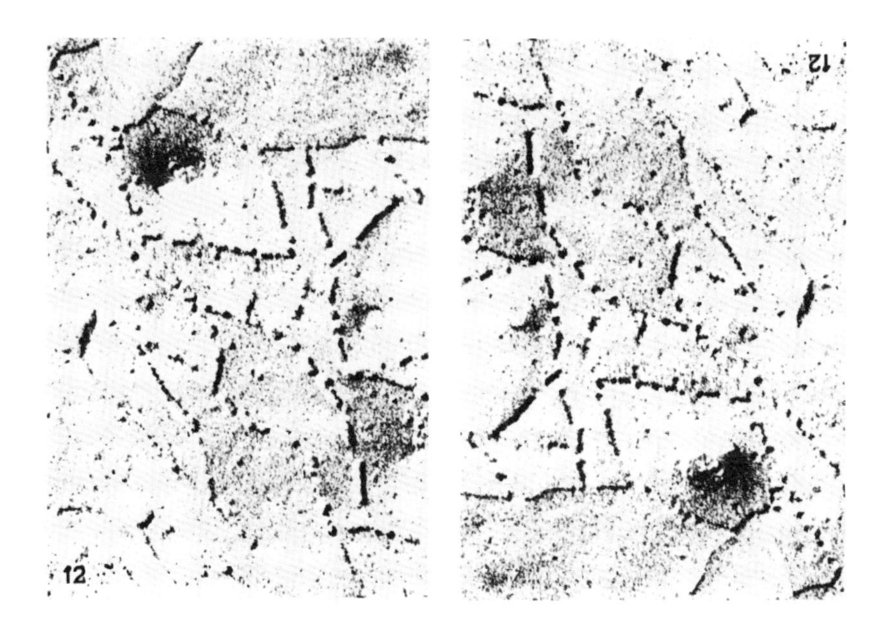

Figure 7

shadow cast by a low beam of light. A close comparison shows I have presented the same photograph twice; at left oriented as it appears in the article, and at right upside down. The inversion of relief arises from a habit fixed in our minds by the canons of classical painting, in which relief is rendered by the shading produced by a source of light apparently situated outside the frame, at upper left.

The purpose of this article is to make the structure of a junction between two cells visible, the two membranes having been cut across obliquely by a technique called freeze fracturing. The photographs are taken with an electron microscope; we are supposed to see "7 to 8 alignments of particles, with numerous intersections, enclosing regions of variable form."[7] The authors, in orienting the photograph, have been careful to respect the "classic" position of the light source, so that the reader "sees" particles and not holes. To do justice to this photograph, I would add that it also is part of a series that is deployed both temporally, showing the growing complexity of the junction (successive additions and new folds, visible as alignments of particles, on the plan of the cut) and spatially, according to the principle of progressive focusing of attention. I have, for the purpose of my demonstration, chosen one of the last of this series, at a very high magnification, where the authors have added neither letters nor arrows, so that I could present it upside down without the manipulation being immediately apparent.

The rules of representation in perspective are another means of rendering depth to which classical painting has accustomed us. Photography tends to make us accept this representation as "natural." However, except in the case of some explanatory diagrams (such as Figures 14 or 19), perspective is rarely used in photographic illustrations in scientific articles. On the other hand, one does find the effect of relief evoked by stereoscope: the same object is represented twice, photographed from a slightly different angle. Special glasses make the left eye look at the left photo and the right eye look at the right, in order to make one "see" the relief (a reconstruction due to the superposition of images in the brain). I offer Figure 8 as an example; unfortunately it is the diagram of a molecule[8] because I have not found any photographs treated in this fashion (If you don't have special glasses, you need only hold a piece of cardboard between the two images, put your nose on the edge of the cardboard—and concentrate a bit).

I would also like to single out an extremely seductive method which, by its very nature, is not reproducible in a journal. This method, used by A. Rambourg on thick sections observed with very powerful electron microscopes, allows the visualization of the "true" relief of a cell as if one were there. The development of the special photographic plate that allows such effects is directly inspired by the "integral photography" of Lippmann:[9] when the viewer moves in relation to the photograph, the details hidden behind the foreground reveal themselves, as they would in reality. In this way it can be seen that certain cellular organs are organized as a system of interconnected tubes.

A "scientific" photograph (one that is publishable in an article) is the complete opposite of a father's photograph of the baby, the

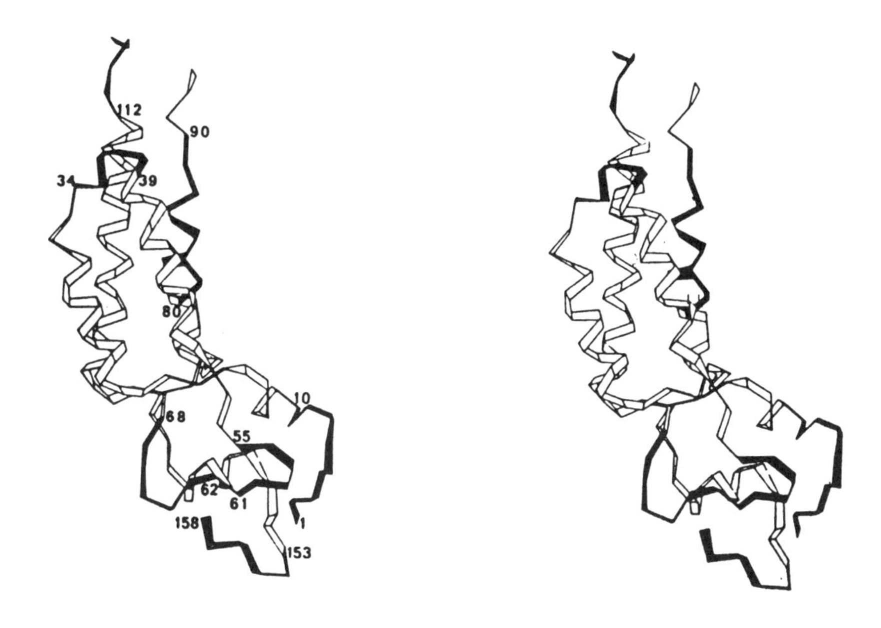

Figure 8

cat, the countryside, and perhaps a bike. . . . The example of Figure 4 shows that it must contain the least information possible, for fear of confusing the meaning. In making the photo part of a series, as in Figure 5 and 6, the uncertainty about the meaning is reduced. The comparison allows one to eliminate everything that doesn't change, and to channel the reader's gaze toward the differences that are the only pertinent details. This process progressively highlights, in the course of the reading of the plate, "what there is to see." The reading of the signification of a photograph (with scarcely any relation to its aesthetic character) relies on the use of systems that semiotics calls "semi-symbolic." They couple a difference at the level of the signified to a marked difference at the level of the signifier; in Figure 1, for example, the light and dark (categories or oppositions belonging to the signifier) are the colors of the background and of the object respectively; the light is the background on which appear the dark spots that make the structure visible. In Figure 3, the difference in shading at the level of the signifier is used with an opposite meaning: the dark background serves to "highlight" the white granules. In Figure 4 there is the same opposition, but it is not meaningful; the darkness indicates at the same time the background and the shadow the object casts on itself. On the other hand, it is necessary to note the difference between the forms (split or nonsplit) of the spokes. In contrast, in Figure 7, the opposition of light and dark serves the spatial dimension, or depth, at the level of the signified.

A spatial opposition at the level of the signifier can be coupled with another spatial dimension at the level of the signified. In paintings represented in classical perspective, the category of the signifier "high vs. low," in the space of the painting, is coupled with the category of the signified "far vs. near," in the eye of the painter or of the spectator. Certain of these oppositions are gradual, like high and low: between the two poles of the upper edge and the lower edge of the painting, all the intermediate points of distance can be inscribed, in a line running to the convergence point at the horizon. Alternatively, distance can be signified by discrete planes (first plane, second plane, etc) where objects of the same size in "reality" can be represented as diminishing in size.

The readability of a scientific document depends on the number of semisymbolic systems put into play; to simplify, we can speak of different "dimensions." Figure 4, though it is rectangular (that is to say two-dimensional), utilizes only one dimension to show information, the opposition between center and periphery. It would mean just as much if one rotated it 90°. In contrast, Figure 3, without going as far as a different definition of high and low and of right and left, as in a map, shows some granules disposed "in elongated ribbons" running in the direction of the fibers. The horizontal presentation of this elongation is, therefore, logical; on the other hand, one could still rotate it 180° without causing any problems. Figure 7 is of the "map" type; its "four cardinal directions" are defined by the positions of a cell of one type at the "northwest" and a cell of another type at the "southeast" (as one will see on the photographs at the lower magnification that precede it). It also uses the third dimension of space, that of distance (even if the effect of relief can be imputed to the lighting). Furthermore, it transfers information through dimensions other than those of space: the position of the elements, their form, their orientation, the state of the surface (rough or smooth). Of all the photos I have shown, it is the one that puts into play the largest number of dimensions. One could add, though I have not given any examples of them, color and transparency as other dimensions that permit the establishment of distinctive traits at the level of the signified. Nevertheless, thanks to the force of habit, these correlations seem "natural" enough.

2.3 Arbitrary space

However, the most frequent case in scientific illustrations is the use of a spatial dimension at the level of the signifier for the deployment of a nonspatial opposition: hot and cold, for example, are often displayed on a vertical scale. There are devices (the thermometer is a good example) that were conceived specifically for this function. One sees more and more illustrations that are photographs of "traces" produced by automatic techniques that display results in a two-dimensional form. Figure 9, which constitutes an enlargement of the last part of Figure 5, is typical enough from this point of view. The legend indicates that it deals with a typical spectrograph obtained by microanalysis, using X-rays, of granules treated with oxygenated water. The abcissa shows, from left to right, the growing energy of the beam of X-rays; a specialist familiar with the corresponding energies would not contest the identifications of peaks with the elements indicated by their chemical symbols. The spatialization from left to right corresponds to a sort of "scan" (as in television). The television tube beam introduces time as a circuit of the image; here also (even though there was only one "line"), it is the time it takes for the display to run through the scale of energies that is implicitly present underneath the representation on a horizontal scale. The ordinate, a specialist would say, is the "height" of the corresponding signal. As in the case of the thermometer, a quantity (here, the intensity of the signal) is spatialized in the vertical dimension.

This photograph is of the "map" type, in that high and low, left and right, are differentiated. Nevertheless, the two-dimensional space used has nothing "realistic" about it. It is completely arbi-

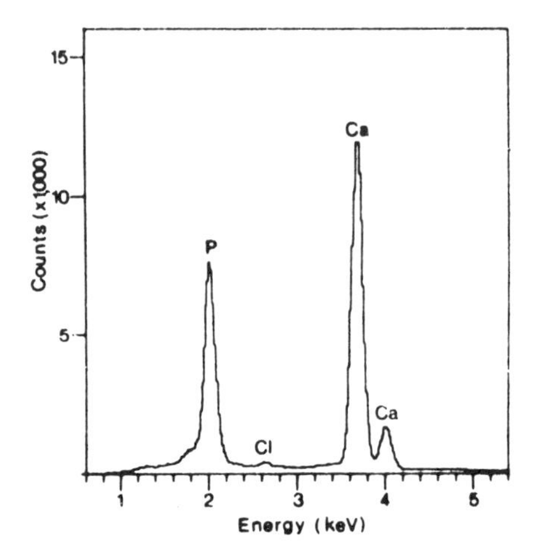

Figure 9

trary and depends on the technique of visualization, which defines the abcissa as a scale of energy and the ordinate as an intensity of signal. In fact, the two scales could display any quantity. "Color," reduced to the opposition of black and white, is used as in a line drawing to mark the limits of the object on the background. Interpretation suggests that such traces are comparable to some cross sections (one speaks of "peaks," as in a relief map), but this does not imply a third dimension of thickness; what matters is the comparison of surfaces delimited by the black outline. In this photograph one sees at a glance the relative abundances of calcium and of phosphorus (indicated by the height of the peaks). However, this figure is not used by the text to identify the nature of the granules; it directs one to a table (Table 1), to affirm that the granules are composed principally of "hydroxyapatite . . . (Ca/P = 1.67), possibly with an admixture of tricalcic phosphate . . . (Ca/P = 1.5)." Then, in reading the legend of this table, one finds that the figures shown there were obtained with other methods of measurement: flame photometry for metals, and a colorimetric method for the

Table 1. Composition of *Nephtys* muscle granules

	Concentration (% of ash) Soluene-treated	H_2O_2-treated
Ca	34.46	36.38
P	16.42	17.49
PO_4	50.34	53.64
Mg	1.75	0.275
Sr	0.0569	0.0514
Ca/P atom ratio	1.62	1.61

phosphate. The photograph duplicates the table in part, but the table could not exist other than in the form of an aggregation of figures, since it is the sole means of assembling the results given by two different techniques.

We must now explain this surprising phenomenon, the existence of equipment that directly makes analogues of the graphics that the researcher could just as well trace with the aid of figures that he can read on the tapes of the counters or the graduations of the instruments. To explain, we must look at the advertising of vendors of such equipment. One advertisement in particular appears revealing to me: "Finally, hands off microtitration."[10] It deals with a programmable device, which has the advantage of being automatic: "It eliminates the tedious manual labor that creates so much potential for error in the course of the pipetting that the researcher requires."

By analogy, a graph produced automatically is done with neither the hands nor the eyes of the experimenter. If we return to the diagram of the technique of visualization presented at the beginning, we see that it is a matter of eliminating as many steps as possible between the object produced by nature and the object transmitted in the article, and especially as many steps as possible where a human subject operator would intervene. It is certainly not a matter of fraud; as the article by Paradies shows, it is easy enough to substitute one photograph for another. It is a matter of the errors that could be introduced through human fallibility: an error in transcribing the figures furnished by the apparatus, confusion of samples, clumsiness. . . . Tables and graphs are constructed by authors; they do not have the authentic quality of a photograph, a recording furnished by an apparatus that offers the reader a guarantee against the intervention of the author.

Photography permits, in principle, the deployment of many dimensions of reading but paradoxically, despite the wealth of possibilities of half-tones and of differentiation of surface textures, these possibilities are not explored in scientific photography. When half-tones and textures are used, the construction of a series eliminates any dimensions that are not pertinent to the process of interpretation. In fact, the reading becomes more difficult as the variations become more gradual. Thus the "demonstration" is as convincing as the contrast is strong. "All or nothing" is preferable to degrees! One can see an illustration of this constraint in the fact that the "map" type images (maps showing localization of radioactive products in the brain, maps of the sky) where the intensity of the light would be "naturally" reproduced with the help of graduations, are seen in "false colors" when they are produced by a computer. Continuous variation is replaced by a scale composed of discrete steps, rendered by colors that are easily distinguished from one another.

2.4 Convergences

All equipment and all experimental conditions can be viewed, as we have seen above, as a technique or a chain of techniques, the function of which is to demonstrate what would not be visible otherwise. In the case of Figure 1, the technique appears at first glance not to correspond to the photographic plate, which functions to preserve the memory of what the technique visualizes. The receiver could be a television screen or the image could be memorized bit by bit by ciphers in the memory of a computer. On the other hand, in Figures 3 and 4 the work of visualization (slicing the tissue, staining, etc.) that surrounds the published photograph seems innocent enough: it aims to show "what is." Of course, it seems easier to show an "object" (such as the skeletons of worms or of sponges) than an *action (un faire)*. In the case of an action, one can show only the result, the *trace*. However, there is not such a sharp distinction between the two types of photographs. In the case of Figure 1, it was also a matter (it's scarcely a metaphor) of visualizing a skeleton. If Figures 3 and 4 (and even 7) appear more "clear," we have already seen that this was an illusion due to our habit of looking at objects in three dimensions in black and white photographs, which reduce space to a plane, and relief to some shapes and some cast shadows. The isolated photograph takes the meaning indicated by the text only through comparison, which allows the elimination of elements that are not pertinent. On the other hand, Figure 1 showed itself quite meaningful for the accusers of the author who published it. The signification of a photo isolated from the article appears most often only to a specialist in the same field. A researcher, even in a closely related field, can evaluate the photo only by reading the description of the technique of visualization (in the "Materials and Methods" section), and the description in the "legend," as in the case of Figure 9 where the scales deploy in two-dimensional space some quantities which have nothing spatial about them.

However, even the presentation of an "object" assumes one does some work on reality, work one can summarize under the headings of selective and contrastive functions, work that also takes place in the case of photography. One can refine the analysis of what I call "the channeling of meaning" by distinguishing two processes that come into play in a photograph, and make it an object *constructed* so that one generally takes it as nature itself. The vocabulary of photography is the source for the terms used to define the operations performed by the technique of visualization, whether it be a simple magnification of the eye, as in a microscope, or constituted of a chain of manipulations devoted to the visualization of an action.

The selective function consists of narrowing the field in order to eliminate as many nonpertinent elements as possible (those elements that make a sort of "background noise" around the image), while still showing the object "whole," without cutting off any piece of it; we will call it "framing." If the emphasis is put on the construction of the frame (the process rather than the result) one can also speak of "focusing". Figure 5 is a good example of this concept: it shows successive, tighter and tighter frames that intervene in the presentation of the granules of calcium phosphate. The contrastive function works within the chosen frame: it consists of cleaning away the nonpertinent elements from the image. Literally, it is an procedure aimed at obtaining an image in which the object is set out sharply against the background; in photography, one can obtain it by appropriate lighting, narrow depth of field that blurs whatever isn't at the right distance, and so on. If one puts the emphasis on the process and not the result, one can also describe this function as "putting in perspective." Figure 6 shows an example of this procedure; the perturbing effect of germanium on the mechanism of the deposit of silicon in the spicules is put into perspective by normal images obtained in its absence, at the top of the page, on another level.

The functions of framing and of contrast affect the *signification*. In the semiological definition, signification is not the same thing as what one might call "meaning," that which is intuitively perceived by the reader/hearer or that which is meant by the author. These two real cases are outside the field of semiotic analysis. On the other hand, semiology understands signification as a process that is constructed and, to some degree, open to evaluation: the amount of signification present in an image depends on the number of distinctions put into play. It depends, then, on the number of semisymbolic systems exploited, and on whether they are gradual or discrete. The use of framing and contrast in the production of scientific images aims to generalize and reduce the signification to the single "message" given in the scientific text it accompanies, which can be considered in most cases as its commentary. I will discuss later the explanatory schema that would be, in contrast, a commentary on the text. There could be no question, in a scientific image, of the "proliferation of meanings" dear to artists, a proliferation one rediscovers in the illustrations of popular science articles. In fact, scientific illustration behaves exactly like the scientific article itself: it is constructed like some military strategy, an ambush without an escape route. Each time that a reading of the results different from that of the authors could possibly be made, the departure is barred by an adequate argument.

I will take as an example the article by Gibbs and Bryan (1984) from which Figure 5 is taken, because its message is expressed in familiar terms. It presents two items that seem at first to be equivalent on the level of information: a table (Table 1) and a graph (Figure 9) that both indicate that the granules are constituted of calcium phosphate. This similarity should allow us to show the respective roles of the two kinds of illustrations in the argument. The article begins with the question of the exceptional chemical composition of nephthys, compared to other marine worms: they contain in their body a higher percentage of calcium and of phosphate than do their "cousins." This chemical beginning appears paradoxical, because the article aims to make one "see" this excess phosphate and calcium in the form of granules in their muscles. But a third of the way into the article, we are told that in 1868 Claparède, observing in an optical microscope these whitish bands in the muscles, had concluded that they were fatty deposits. Had the authors begun with a photograph of the granules, they would have had to present a chemical analysis of the fat content of the muscles! This first decision between alternatives, between a method of optical analysis and a method of chemical analysis, is delayed until the authors, having presented a table of data establishing the significant different in the level of calcium phosphate (and not of fat) between nephthys and its cousins, can then present some photographic results. We have noted that the last photo of the series, a spectral analysis showing that the granules are composed of calcium phosphate, is not discussed in the text, which refers to Table 1. This table, which also shows measurements of other metals, gives precise figures for the identification of the chemical nature of the granules. The agreement of the respective quantities of calcium and of phosphate is explicit (although it could be evaluated in only a rough fashion by the agreement of curves in Figure 9). Therefore the function of this photo is to block the route of a reader who, looking only at the photographs, would return to the old interpretation and would "see" globules of fat. The text, on the other hand, uses persuasion to exploit the quantitative results.

The authors must furnish a justification for the presence of these grains (if they were taken as fat, the interpretation of them as a reserve of calories would be "self evident"). They must then establish a link between the presence of these granules and the superior capacity of the worm to bury itself in sediments efficiently and rapidly. Two alternatives concerning the possible function of the granules are presented consecutively: (1) It could be a place for the organism to store calcium, (2) it could be a means of detoxification for the organism (a sort of internal garbage can). They respond to these two possibilities with a double argument that disassociates the presence of the granules from the worm's conditions of life. The topographical approach—worms taken from an environment rich in toxic heavy metals do not have more granules, and the composition of the granules stays the same—is set alongside a temporal approach—the young worms have as many (and no fewer) granules, relative to their size, as the old ones. The alternatives that are closed off in this way make the article a continuous line, leading inescapably to the conclusion: "No comparable accumulation of calcium phosphate in muscle cells is known in the animal kingdom. We suggest that a novel flexible skeleton has evolved in *nephtys* as an integral part of its highly successful burrowing mechanism."

If scientific illustration aims at the same absence of freedom in possible readings as the text, one might ask why numbers, with their one dimensionality and perfect gradualness, are not preferred to photography—why articles continue to present information in two forms (table and photograph), as in the case we have just seen. One might think, following the accusation of Paradies, that "numbers," what Lynch calls *mathematicization,*[11] are a more precise sort of memory than photographs, since it was in a data bank gathering the measurements of different crystals that the identity of the photographed molecule was recognized. However, I do not agree with Lynch's view that mathematicization is the ultimate end; it seems to me it is a back-up plan for when visualization does not furnish an image that is directly interpretable. One senses the regret of the researchers Lynch studies when they explain, in the legend he reproduces, that they had to resort to putting neurons in a chart only because the difference they were looking for was not immediately perceptible.

From these examples, we have been led to distinguish the processes of visualization, strictly defined (which are tied to the mechanism of translation used and which furnish part of the information necessary for the attribution of a signification to the image) from the processes of interpretation of the image itself in which a large part is played by the legend and indications, often in the form of letters or arrows added over the photograph itself. But the technique of visualization used constrains the possible representations of the results. It is necessary to distinguish narrow-aperture techniques from wide-aperture techniques (in my metaphor, I move from photographic equipment to spectroscopes). I mean that there are some techniques that allow no latitude at all for showing phenomena that were not defined in advance and thus not expected: the proportion of phosphate used to fill out the corresponding figures of Table 2 says nothing about the calcium. The measurement of phosphate displays only one dimension: that of quantity. So the only way of presenting the results of proportions of calcium and phosphate together is through the numerical expressions of quantities. The same is true when it is a question of making a synthesis of several identical experiments, as we will see. Even the technique that furnishes the quantities of calcium and of phosphate at the same time, in the form of a record that can be photographed (Figure 8), is of the "narrow slit" type. It can use only two dimensions of space, the vertical and the horizontal, to project two magnitudes. This is what limits the quantity of information that can be transmitted by these automatically traced graphs. Optical techniques, on the other hand, leave room for the unexpected in the sense that they allow, at the same time on the same image, the use of several variables: form, relative disposition of elements, "texture" of the surface. This is true even of the differential transparency of electrons to the electron microscope, which by a technique known as "shading" can be exploited to produce an effect of relief, or make it seem "as if" the object was detached from the background.

3. Difficult choices

3.1 The fourth dimension

The choices made by experimenters are varied, and sometimes contradictory. This is shown by the example of the granules of the nephthys, where the article has both a table, which allows the display of all the results obtained by several methods at several different moments, and the automatically produced graph from what the researchers call generally "a representative experiment." But the illustration, whatever its nature, is constituted in an argument without a response: it constitutes an internal referent (a reality), to which the text points to "show" the discovery of which it speaks.

Photography reduces reality to the bidimensionality of paper: it is a flattened "reality," easy to classify and highlight, if necessary, easily communicable, and lending itself well to comparisons with

Table 2

		C	J	MR1	MR2	MR3	MW1	MW2	BP	PL
Experiment, July 17, 1959	HTO i.p.m./mg fresh tissue $\times 10^{-2}$	31.5	13.2	4.5	2.2	1.8	2.0	2.0	1.9	
Hamster (96 g)	^{23}Na μg/mg fresh tissue	1.4	2.6	3.4	4.4	5.4	8.3	7.4	8.9	
Time elapsed: 0.5 min.	^{22}Na i.p.m./mg fresh tissue $\times 10^{-2}$	3.9	5.1	4.9	4.5	5.2	6.1	2.4	0.7	
Experiment, July 31, 1959	HTO i.p.m./mg fresh tissue $\times 10^{-2}$	17.1	16.9	11.3	6.8	4.8	2.5	3.1	2.2	12.2
Hamster (108 g)	^{23}Na μg/mg fresh tissue	1.7	2.3	3.1	3.8	4.3	5.6	6.2	7.8	3.1
Time elapsed: 1 min.	^{22}Na i.p.m./mg fresh tissue $\times 10^{-2}$	2.7	4.2	5.9	7.1	7.5	9.6	9.6	8.9	3.8
Experiment, July 10, 1959	HTO i.p.m./mg fresh tissue $\times 10^{-2}$	15.8	16.2	13.1	9.4	5.4	4.7	3.9	3.3	9.7
Hamster (110 g)	^{23}Na μg/mg fresh tissue	1.7	2.7	3.4	4.6	5.7	7.9	8.1	8.7	
Time elapsed: 2 min.	^{22}Na i.p.m./mg fresh tissue $\times 10^{-2}$	2.1	3.6	5.0	6.3	8.4	10.5	10.7	11.5	3.1
Experiment, July 16, 1959	HTO i.p.m./mg fresh tissue $\times 10^{-2}$	6.8	7.0	7.3	6.3	5.9	4.4	4.2	3.9	7.6
Hamster (96 g)	^{23}Na μg/mg fresh tissue	1.4	2.8	3.1	4.1	4.8	5.9	6.7	7.0	3.7
Time elapsed: 6 min.	^{22}Na i.p.m./mg fresh tissue $\times 10^{-2}$	1.3	2.3	3.2	4.2	5.3	6.2	7.6	8.2	1.9
Experiment, July 31, 1959	HTO i.p.m./mg fresh tissue $\times 10^{-2}$	7.1	7.1	7.7	6.9	7.2	6.5	6.1	5.7	6.2
Hamster (120 g)	^{23}Na μg/mg fresh tissue	1.7	2.7	3.3	4.1	5.4	5.3	6.0	7.7	3.4
Time elapsed: 10 min.	^{22}Na i.p.m./mg fresh tissue $\times 10^{-2}$	1.1	1.6	2.3	2.9	3.6	3.9	4.4	5.4	2.1

C: Cortex. J: Cortico-medullar junction. MR1, MR2, MR3: Red medulla (outer). MW1, MW2: White medulla (inner). BP: Base papilla. PL: Plasma.

other photographs. It is therefore a nearly ideal form of memorization: with a capacity much superior to that of the "bit," it allows a specialist to grasp the "facts" at a glance instead of relying on the intervention of sophisticated decoding equipment. We have seen, moreover, that though the reality that appears there is constructed, it takes on a role of a guarantee because of its automatic production. However, photography has not replaced all other types of illustration in scientific articles.

There are two forms of illustration that profit both from the advantages of number and the possibility of exploiting two dimensions simultaneously: these are the table and the graph, both of which are organized in two dimensions in the form of a rectangle on a piece of paper, as flat and easy to compare as photographs. Mathematicization seems to be useful essentially in results of the gradual type, since we have seen that scientific photographs prefer contrast to shadings of black and white. Gradualness is "rendered" well in a graph, where it finds itself translated uniformly into the two dimensions of high/low and right/left. The "background," which can occupy a considerable surface area compared to the "object" in a photograph like Figure 4, is reduced to a simple line of coordinates, while gradualness is represented on the scale. The number of dimensions is limited only by the number of distinct symbols that one can make fit in the rectangle of coordinates. In a table, the "background" is again reduced: it is the figure zero that it is not even necessary to mention. The number of dimensions is almost unlimited: it can have as many of them as there are rows and columns in the table. As one can write out ("title") each one, one does not even have to limit oneself to some symbols that one can differentiate easily (black or empty circles, squares, triangles, etc.).

The graph supposes a calculation preceding it: it constitutes, in fact, the presentation of a table of figures in another form. One might ask why one sees graphs more often than tables in scientific articles. What are the advantages, respectively, of the table and the graph? It seems that it is a problem of readability: the graph can be interpreted at a glance, while the table asks one to compare figures and mentally make some subtractions, additions, multiplications, and divisions. The graph is probably more "convincing" because it economizes the time and the attention of the reader. If we compare it to a photograph, we see that a graph can support many more dimensions than a photograph while remaining readable. Furthermore, the graph allows one to represent "the fourth dimension," that of time, translating it into a spatial dimension. Time, in a graph, unfolds conventionally from left to right (perhaps like reading in the West). For someone used to reading it it introduces a dynamic: it replaces the trace of the action as a static "image," in which the reader reconstitutes imaginatively what could have happened before and what could result from it, with the effect of the unfolding of the phenomenon as if under the eyes of the observer.

Table 2 is a perfectly undecipherable table that attempts to demonstrate the water-circulation mechanism through countercurrent in a hamster kidney.[12] The first two columns are used to give "meaning" to the numbers of the other two columns; the titles of the columns, composed of letters and of indexes, correspond to "pieces" of the kidney distinguished by visible anatomical criteria. Figure 10, taken from the same article, gives the same information in the form of a "comic strip": in each frame, the time runs from zero to ten minutes, on the abcissa; on the ordinate, the quantity of the radioactive indicator measured is expressed in a percentage of the value in the cortex (the outermost part of the kidney). One of the columns of the table is replaced by a dotted line at the level of 100%. This technique makes it possible to give the same scale for variations of intensity of the sodium indicator and the water indicator: they are

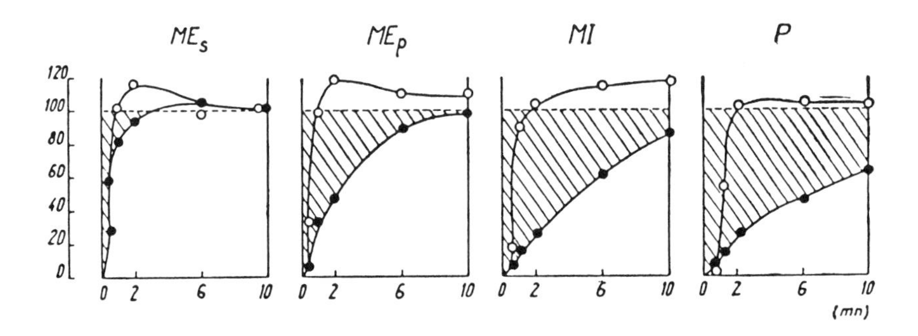

Figure 10

both represented, water with filled circles and sodium with open circles. The demonstration of the counter current, as in the table, rests on the comparison of how the two develop, but the juxtaposition of the curves makes the comparison on the graph immediate. Each of the four frames dramatizes what happens in a different level of the kidney, from the surface (at left) to the core (at right); each of the frames, except the last, regroups the indications of two columns. I use the term *dramatizes* intentionally, because the curve that links the circles reestablishes a sort of continuity of the phenomenon, while the measures in time are discrete and, according to the information on the table, were each made on a different animal. The curve even gives a value at time zero, when no measure was made. The sodium/water comparison is redoubled with a shaded area which, the legend says, "indicates the slowness of the renewal of water compared to its renewal in the cortex." It is this area, which grows from left to right, that makes visible, in the four frames taken together, the phenomenon of exchange of water by countercurrent.

3.2 Continuity and diversity

Figure 11 illustrates the same phenomenon, in a rabbit, this time in the form of a juxtaposition of three photographs.[13] The one in the middle is an anatomical photograph, showing the cut made in the kidney to perform the experiment; the first is an autoradiograph of this same section, where the radioactive water, injected in the animal as an indicator, is made visible through the effect of its radioactivity on a photographic plate wrapped in paper. The third picture shows the same thing for radioactive sodium. This series accounts for only one of the cases on the graph, that corresponding to the time of two minutes on the graph for the hamster, where the difference in the distribution of radioactivity of water and of sodium is the greatest (for the rabbit, the time is five minutes).

This photographic montage was published in a later article, a review article, where the preceding graph was also published. It is interesting to note that in this article, the table is no longer used. The authors' choice makes one think that the photographic series illustrates more "effectively" the slowing of the renewal of water

Figure 11

compared to that of sodium in the core of the kidney, probably because it shows this core in place, in the automatic continuity of the kidney, while the graph juxtaposes (in order) four levels, so that the observer can reconstruct the continuity himself. Furthermore, the time chosen is that which makes the contrast stronger. The core is white for the water, black for sodium. Though it contains less information it can, however, be said to be more convincing. The graph sets out its information in four dimensions: time (from 0.5 to 10 minutes) intensity of radioactivity (this and the time are presented gradually, on the abcissa and the ordinate of each frame), topography from the surface of the kidney to its core (in the four frames, the surface [cortex] represents 100), and the comparative dimension between water and sodium (categorically, by the juxtaposition of two sets of symbolism). The series of photographs, on the other hand, uses only comparison, topography, and, for the intensity of radioactivity, black and a white that are scarcely shaded. The graph therefore allows the use of many more dimensions than photography; it approaches, in its "comic strip" construction, the series of photographs that we described earlier in the discussion of the channeling of meaning. Figure 12, however, presents information collected in a single graph, using the distinction between the shaded and nonshaded sections to distinguish two different times of incubation; the ordinate sets out the intensity varying with height, and the abcissa sets out a topography.

This graph also uses the tracing of a continuous curve to "render" the anatomic continuity of the kidney from its surface to its core. Figure 12 takes the form of a "bar graph."[14] The intensity, proportional to the height of the "bar," represents the activity of an enzyme administered in several different segments of the nephron (a nephron is the most basic functioning unit of the kidney). It is a matter, then, of dramatizing an "action": the trace of this action, the quantity of substrate hydrolyzed by an individual segment (only a figure given by the counter) has disappeared in the different normalization calculations based on the controls, the "blanks," the length of the tube, and the mean of the various runs. Each bar is equivalent to a column of a table. If this table gave details on all the measures performed on each type of segment, one would also find in it, probably in the last line, the mean and the statistical deviation from the mean, which would allow one to compare each segment to

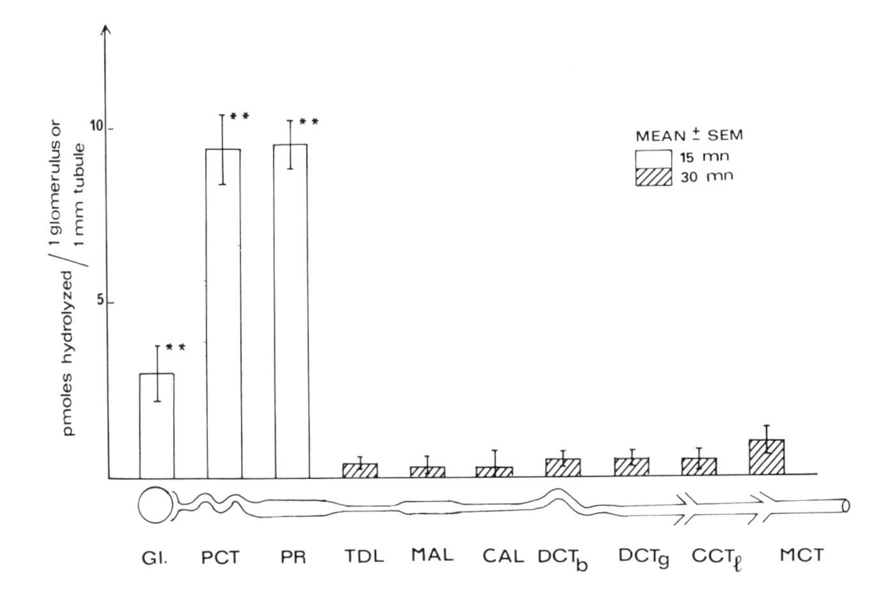

Figure 12

the following one. On the graph, the height of the bar represents the mean of the measures and the figure printed above shows the number of rabbits used (each one in the course of a different experiment). The result of a statistical analysis is represented by the length of the vertical line cutting across the top of each bar; the presence of stars signals differences that are "significant" (the number of stars corresponds to the "threshold"). As in the previous case, the treatment is performed on "bits," cutting the core of the kidney into discrete units. Continuity is partially reestablished on the axis of the abcissa, where the segments are arranged in order. But, more especially, the abcissa is paralleled by the little diagram that is paired with the graph, representing an uncoiled nephron.

3.3 The explanatory diagram

The explanatory diagram or schema is another means of restricting the number of possible interpretations of an image, but it is a direct means, one that reveals the interpretation of the author. We have seen how the reading of the graph in Figure 12 was enlivened with a diagram of a nephron giving meaning to the abcissa, and acting as the double of the conventional abbreviated names printed below the "bars." The function of this schematic representation was also to put the dissected bits of the nephrons—the whole of which were represented by the bases of the bars, with their constant lengths and regular spacing—back together in the anatomic continuity of the structure. Its function is to show what one cannot see: the relation of continuity between the elements, which the diagram breaks down, so that one can also compare the activity of the enzyme from one segment to another. Other diagrams serve the function of softening the static character of the image by introducing what one could call movement. In Figure 13, which represents the operation

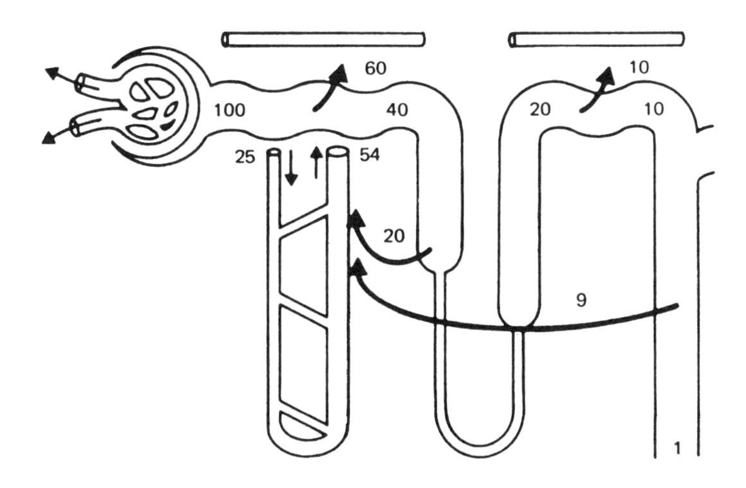

Figure 13

of the nephron in the course of concentrating urine, the horizontal passage of water is represented by the arrows.[15] In this diagram, the nephron is conventionally oriented as in a cross section, where the cortex would be above and the papillae are below (that is, as if the photos presented in Figure 11 were turned upside down). Thus, the reader is warned that the water does not follow the hairpins of the canal but takes a short cut that slows the renewal of the water in the interior regions of the kidney. The usefulness of a diagram for visualizing action is undeniable: without this artifice, another medium of presentation, such as a film, would be necessary. In a two-dimensional medium one can see only the trace, the result of the action, or the "object." Other diagrams, like that of Figure 19, present the organization of an object in three spatial dimensions.

I will not enlarge on the relationship of diagrams and photographs, a point treated in detail by Lynch; I will just give one humorous example. Figure 14, placed at the beginning of an article,[16] serves as an explanation of figure 15, which is found in the middle of the article; it could also, if it represented two cells instead of one, serve as a commentary on the freeze fracture given in Figure 6. In fact, the presence of this type of schema is much more common in popular science articles than in specialized articles. Here, the action is represented by some characters dressed warmly and equipped with picks. In a more "serious" diagram, the action is represented by arrows, and the Subject-Operator (when this is not the experimenter himself) is often designated by its chemical symbol or a more or less conventional abbreviation placed near the arrow.

3.4 "Linguistic" transcriptions

One sees more and more often in genetics articles images that are halfway between text and a diagram. Figure 16 shows various sequences of the gene of the protease inhibitor; the gene from the mouse is used for comparison with that of primates.[17] The linear connection of the elements of the sequence of DNA is shown by the succession of letters indicating the nucleotides with the initial letter of the base, and by the numbering (from the left of each row). But it does not show a double helix; the complementary chain is

Figure 14

not represented. Neither is the shape in space. So is this a diagram or a printed text? As in a text, one must proceed line by line, from left to right, then from top to bottom. There are some "blanks" where dashes replace the letters; they are, the legend says, "some regions where additions or deletions of nucleotides appear." One also sees a boxed section that corresponds to "the ten amino acids of the active centre" of the molecule. The reading of this image requires a certain amount of mental gymnastics. The legend makes us pass without any transition from the gene to the protein it produces; the two are actually visualized simultaneously. The nucleo-

Figure 15

tide sequence is "translated," above for the mouse, below for the primate, by a letter-code symbolizing one of the twenty amino acids that make up the proteins. The presence of this letter transforms the continuous chain of initials of bases by articulating them implicitly into "triplets," groups of three letters representing the particular combination of nucleotides that "codes" for one of the amino acids. However, the complementary chain of the one that is written out, though it is absent here, is not absent from the commentary: "the dots," in fact, "indicate nucleotides which are homologous in

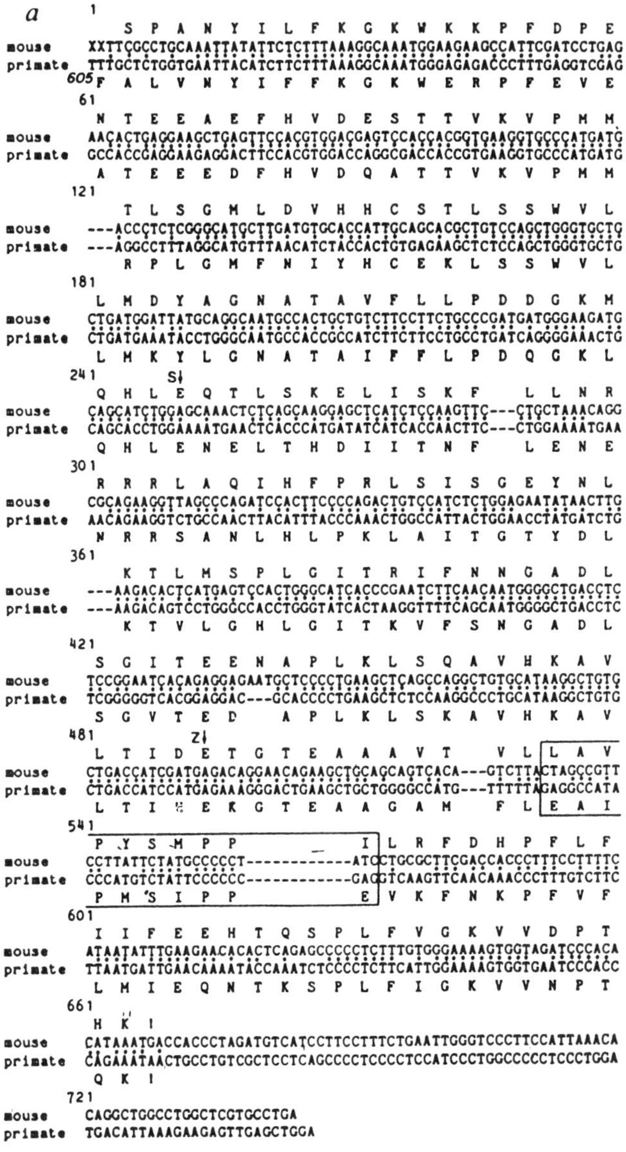

```
a      1
          S  P  A  N  Y  I  L  F  K  G  K  W  K  K  P  F  D  P  E
mouse     XXTTCGCCTGCAAATTATATTCTCTTTAAAGGCAAATGGAAGAAGCCATTCGATCCTGAG
primate   TTTGCTCTGGTGAATTACATCTTCTTTAAAGGCAAATGGGAGAGACCCTTTGAGGTCGAG
     605 F  A  L  V  N  Y  I  F  F  K  G  K  W  E  R  P  F  E  V  E
      61
         N  T  E  E  A  E  F  H  V  D  E  S  T  T  V  K  V  P  M  M
mouse     AACACTGAGGAAGCTGAGTTCCACGTGGACGAGTCCACCACGGTGAAGGTGCCCATGATG
primate   GCCACCGAGGAAGAGGACTTCCACGTGGACCAGGCGACCACCGTGAAGGTGCCCATGATG
         A  T  E  E  E  D  F  H  V  D  Q  A  T  T  V  K  V  P  M  M
     121
            T  L  S  G  M  L  D  V  H  H  C  S  T  L  S  S  W  V  L
mouse     ---ACCCTCTCGGGCATGCTTGATGTGCACCATTGCAGCACGCTGTCCAGCTGGGTGCTG
primate   ---AGGCCTTTAGGCATGTTTAACATCTACCACTGTGAGAAGCTCTCCAGCTGGGTGCTG
            R  P  L  G  M  F  N  I  Y  H  C  E  K  L  S  S  W  V  L
     181
         L  M  D  Y  A  G  N  A  T  A  V  F  L  L  P  D  D  G  K  M
mouse     CTGATGGATTATGCAGGCAATGCCACTGCTGTCTTCCTTCTGCCCGATGATGGGAAGATG
primate   CTGATGAAATACCTGGGCAATGCCACCGCCATCTTCTTCCTGCCTGATCAGGGGAAACTG
         L  M  K  Y  L  G  N  A  T  A  I  F  F  L  P  D  Q  G  K  L
     241        S↓
         Q  H  L  E  Q  T  L  S  K  E  L  I  S  K  F     L  L  N  R
mouse     CAGCATCTGGAGCAAACTCTCAGCAAGGAGCTCATCTCCAAGTTC---CTGCTAAACAGG
primate   CAGCACCTGGAAAATGAACTCACCCATGATATCATCACCAACTTC---CTGGAAAATGAA
         Q  H  L  E  N  E  L  T  H  D  I  I  T  N  F     L  E  N  E
     301
         R  R  R  L  A  Q  I  H  F  P  R  L  S  I  S  G  E  Y  N  L
mouse     CGCAGAAGGTTAGCCCAGATCCACTTCCCCAGACTGTCCATCTCTGGAGAATATAACTTG
primate   AACAGAAGGTCTGCCAACTTACATTTACCCAAACTGGCCATTACTGGAACCTATGATCTG
         N  R  R  S  A  N  L  H  L  P  K  L  A  I  T  G  T  Y  D  L
     361
            K  T  L  M  S  P  L  G  I  T  R  I  F  N  N  G  A  D  L
mouse     ---AAGACACTCATGAGTCCACTGGGCATCACCCGAATCTTCAACAATGGGGCTGACCTC
primate   ---AAGACAGTCCTGGGCCACCTGGGTATCACTAAGGTTTTCAGCAATGGGGCTGACCTC
            K  T  V  L  G  H  L  G  I  T  K  V  F  S  N  G  A  D  L
     421
         S  G  I  T  E  E  N  A  P  L  K  L  S  Q  A  V  H  K  A  V
mouse     TCCGGAATCACAGAGGAGAATGCTCCCCTGAAGCTCAGCCAGGCTGTGCATAAGGCTGTG
primate   TCGGGGGTCACGGAGGAC---GCACCCCTGAAGCTCTCCAAGGCCCTGCATAAGGCTGTG
         S  G  V  T  E  D     A  P  L  K  L  S  K  A  V  H  K  A  V
     481           Z↓
         L  T  I  D  E  T  G  T  E  A  A  A  V  T     V  L  L  A  V
mouse     CTGACCATCGATGAGACAGGAACAGAAGCTGCAGCAGTCACA---GTCTTACTAGCCGTT
primate   CTGACCATCCATGAGAAAGGGACTGAAGCTGCTGGGGCCATG---TTTTTAGAGGCCATA
         L  T  I  [illegible]  E  K  G  T  E  A  A  G  A  M     F  L  E  A  I
     541
         P  Y  S  M  P  P              I  L  R  F  D  H  P  F  L  F
mouse     CCTTATTCTATGCCCCCT-------------ATCCTGCGCTTCGACCACCCTTTCCTTTTC
primate   CCCATGTCTATTCCCCCC-------------GACGTCAAGTTCAACAAACCCTTTGTCTTC
         P  M  S  I  P  P              E  V  K  F  N  K  P  F  V  F
     601
         I  I  F  E  E  H  T  Q  S  P  L  F  V  G  K  V  V  D  P  T
mouse     ATAATATTTGAAGAACACACTCAGAGCCCCCTCTTTGTGGGAAAAGTGGTAGATCCCACA
primate   TTAATGATTGAACAAAATACCAAATCTCCCCTCTTCATTGGAAAAGTGGTGAATCCCACC
         L  M  I  E  Q  N  T  K  S  P  L  F  I  G  K  V  V  N  P  T
     661
         H  K  !
mouse     CATAAATGACCACCCTAGATGTCATCCTTCCTTTCTGAATTGGGTCCCTTCCATTAAACA
primate   CAGAAATAACTGCCTGTCGCTCCTCAGCCCCTCCCCTCCATCCCTGGCCCCCTCCCTGGA
         Q  K  !
     721
mouse     CAGGCTGGCCTGGCTCGTGCCTGA
primate   TGACATTAAAGAAGAGTTGAGCTGGA
```

Figure 16

the *pairs* of sequences" (my emphasis). It is true that, as the bases in the double helix of DNA are always paired in the same fashion, it is not difficult, for whoever knows the rule, to "see" the absent complementary sequence at the same time as the other. Since it is a question of making the reader see the similarities and the differences between the sequences, and since the sequences are (by definition) constituted by a linear chain, the representation in the form of a "text" to be read appears appropriate enough. The economy with space that allows symbolization by letters of the alphabet (instead of the complete name or, worse yet in terms of crowding, of the chemical formula) seems to be more precise than the actual nucleotides that link to each other, from the fact that even as they lend themselves to a chain and combine with each other, they resemble each other enormously. One sees in Figure 17 a representation of the double helix in its "chemical" form; at left the atoms are represented by clustered balls, while at right the ties between atoms are represented by bars.[18]

Figure 18 presents, flattened out, one of these famous transfer RNAs "in the form of a clover leaf" that Paradies tries to crystallize in order to establish its spatial form.[19] The fact that it is composed of a long chain of ribonucleic acids is indicated by the numbering by tens, but each RNA of the chain is symbolized as previously by the initial of the corresponding base. The fact that the structure is due to the folding of the chain back on itself by base pairing of complementary sequences is indicated by a little dot placed between two letters. This discrete little dot symbolizes several hydrogen bonds and a double helical structure (like that of DNA), but when the bonds are absent, one sees that the letters separate and form some kind of bubble. The representation of the thing is adequate to the message of the article: two little rectangles, below and above right, surround two letters. The column that supplements the figure under each rectangle indicates the various mutations affecting the two bases in question and, at least in the case of the lower rectangle, indicates the change in the meaning of the code that results. One can compare it usefully to the representation in Figure 19 that is taken from a textbook[20] and tries to give an idea of the form in space, with the aid of an entirely different symbolism; the bases are represented with rectangles and squares drawn in perspective, and the hydrogen bonds are materialized with lines be-

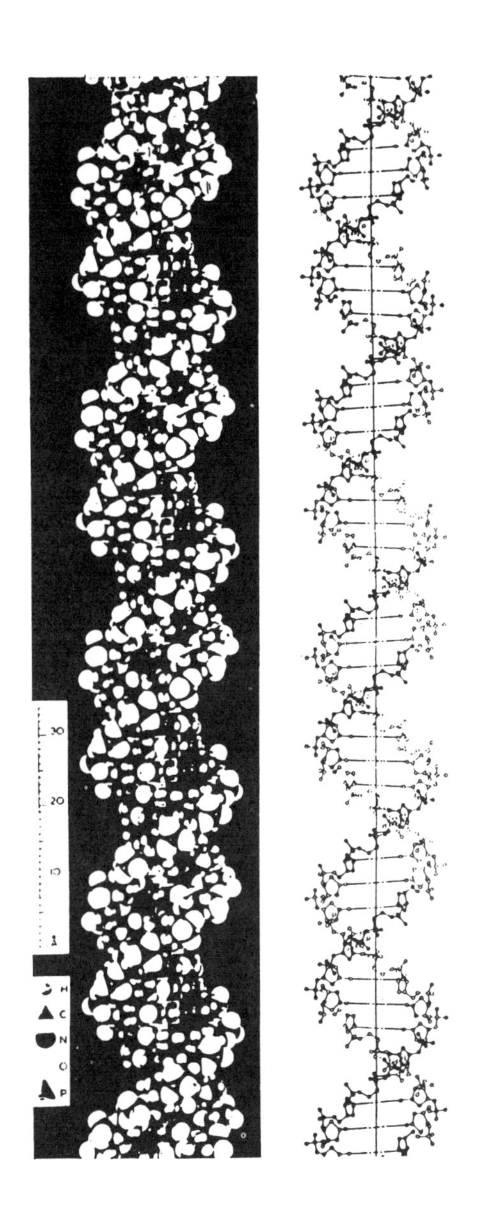

Figure 17

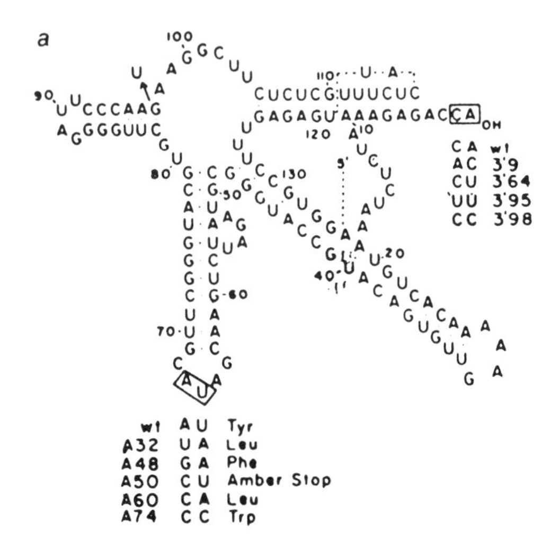

Figure 18

tween the rectangles. The ribose phosphate part of the nucleotide, by which each base is attached to those that precede and follow it in the chain, is symbolized by a continuous ribbon bearing numbers, while it showed its presence in the other figure only in a residual form, by the numbering scheme.

To conclude, I believe neither in mathematicization nor in "linguisticization." It is a matter of persuasion, and every approach is good; some features however work better, such as the economy of place that allows one to see the result at a glance, and automatic recording, which serves as a guarantee of "reality." The biologist shapes these abstractions, the number and the letter, without the image of that of which he speaks ever being absent, as is shown by the double helix implicit under the diagrams made of letters, or the nephron folded in a hairpin under the linear diagram. I will not call this geometricization as Lynch does—this assimilates it with the work done on numbers. It is rather a sort of inveterate materialism. However, the materiality is not that of the "real" world. It is that

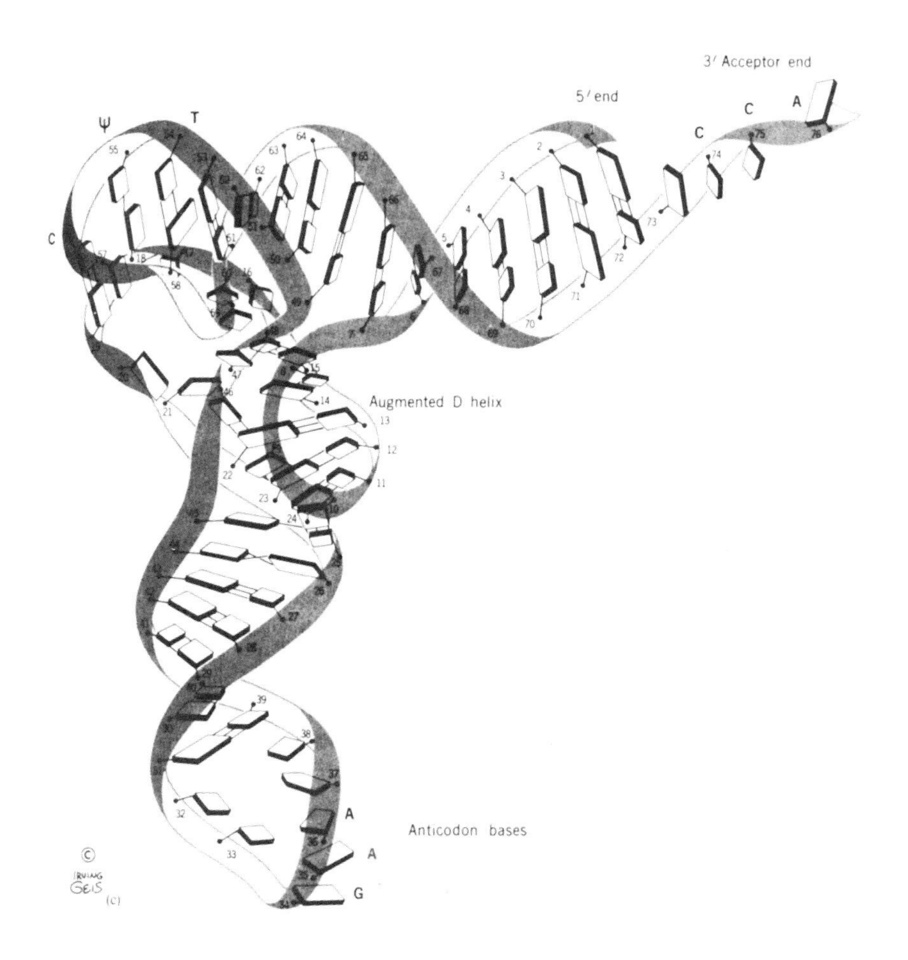
3′ Acceptor end
5′ end
Ψ
T
C
C
C
A
Augmented D helix
Anticodon bases
A
A
G
©
IRVING GEIS
(c)

Figure 19

of a representation that remains schematic: a vision that processes the great variety of the real to make discrete units of it, but which includes movement and the three dimensions of space.

Notes

1. The controversy includes three texts published in *Nature* 303 (1983): in the correspondence, "True identity of a diffraction pattern attributed to valyl tRNA," by W. A. Hendrickson, B. E. Strandberg, A. Liljas, L. M. Amzel, and E. E. Lattaman, p. 195, and the response by H. H. Paradies, "A reply from Paradies," p. 196. In "News and Views," p. 197, there is an article by P. Newmarck, "Disputed X-ray data unresolved," with the subtitle: "The letters from Hendrickson and Paradies on the two preceding pages raise doubts about the authenticity of a series of published papers. Here is a guide for readers."
2. H. H. Paradies and J. Sjöquist (1970), "Crystallographic study of valine tRNA from yeast," *Nature* 226, 159–161.
3. M. Callon (1981), "Traductions et boîtes noires," *Economie et Humanisme*, 53–59.
4. Bruno Latour (1985), "Les vues de l'esprit: visualisation et connaissance scientifique." *Culture Technique* 14.
5. P. E. Gibbs and B. W. Bryan (1984), "Calcium phosphate granules in muscle cells of *Nephtys* (Annelida, Polychaeta)—a novel skeleton?" *Nature* 301, 494–495.
6. T. L. Simpson and P. F. Langenbruch (1984), "Effects of Germanium on the morphogenesis of a complex silica structure and on the assembly of the collagenous gemmule in a freshwater sponge," *Biology of the Cell* 50, 181–190.
7. The photograph is taken from an article by M. Delbos, K. R. Miller, and J.-D. Gipouloux (1984), "Freeze-fracture of *Rana pipiens* gonad anlage: study of primordial germ cells and other cellular types," *Archive d'anatomie microscopique* 73, 57–67. This effect of inversion of relief has been noted previously by D. J. Cook (1983), "Relief in Microscopy," in the correspondence in *Nature* 306, 428.
8. E. Westhof, D. Altshuh, D. Moras, A. C. Bloommer, A. Mondragon, A. Klug, and M. H. V. Van Regenmortel (1984), "Correlation between segmental mobility and the location of antigenic determinants in proteins," *Nature* 311, 123–126.
9. M. G. Lippmann (1908), "Epreuves réversibles. Photographies intégrales," *Comptes rendus de l'Academie des sciences* 146. This method was actually put into practice by A. Rambourg and Bonnet.
10. Advertisement for the Cetus Corporation.
11. M. Lynch, "The externalized retina," this volume.

12. F. Morel, M. Guinnebault, and C. Amiel (1960), "Mise en évidence d'un processus d'échange d'eau par contre-courant dans les régions profonds du rein de hamster," *Helvetian Physiology and Pharmacology,* 18, 183–192.
13. F. Morel and M. Guinnebault (1961), "Les mechanismes de concentration de l'urine," *Journal de Physiologie* 53, 75–130.
14. J. Marchetti (1984), photograph prepared for a "poster session" entitled "Métabolisme de la lysine-bradykinine par les glomérules et les segments microdisséqués du néphron," presented at the meeting of the Institute of Biology of the Collège de France, (provided by the author).
15. Taken from P. Deetjen, J. W. Boylan, and K. Kramer (1978), *Physiologie du rein et de l'équilibre hydroélectrique,* Paris: Masson, 23.
16. S. B. Horowitz, and D. S. Miller (1984), "Solvant properties of ground substance studied by cryomicrodissection and intracellular reference-phase techniques," *The Journal of Cell Biology* 99. 1, 172s–179s.
17. R. E. Hill, P. H. Shaw, P. A. Boyd, H. Baumann, and N. D. Hastie (1984), "Plasma protease inhibitors in mouse and man: divergence within the reactive centre regions." *Nature* 311, 175–177.
18. Nobel Lecture by M. H. F. Wilkins (1962): "Molecular configuration of nucleic acids," in *Nobel Lectures, physiology or medicine, 1942–1962* (Amsterdam-London-New York: Elsevier, 1964), 761.
19. T. W. Dreher, J. J. Bujarski, and C. T. Hall (1984), "Mutant viral RNAs synthesized *in vitro* show altered aminoacylation and replicase template activities," *Nature* 311 pp. 171–175.
20. C. R. Cantor, and P. R. Schimmel (1982), *Biophysical Chemistry* I, "The conformation of biological macromolecules," San Francisco: W. H. Freeman.

Every picture tells a story: Illustrations in E.O. Wilson's *Sociobiology*

GREG MYERS*
Modern Languages Centre, University of Bradford, Bradford, West Yorkshire BD7 1DP, UK

1. Introduction

If a child were to look into *Nature* and, say, *Sociology* or the *Journal of Linguistics*, the first thing that might strike him or her as important would be that the scientific journal had pictures, while the others just had print. Those of us who study scientific texts have, until recently, ignored these pictures.[1] But since Martin Rudwick commented on this lack of attention in 1976, a number of studies of scientific discourse have discussed the use of illustrations in scientists' communications with scientists (Latour, 1985; Shapin, 1984; Lynch, 1985c; Bastide, 1985). Illustrations are also important in communications between scientists and readers outside their specialties (Jacobi, 1985, 1986; Gilbert and Mulkay, 1984; Pickering, 1988). Indeed, the iconography of a science is more likely to have an impact on the public than the words or mathematics, which may be incomprehensible to them. If we ask, for instance, what most people would recognize from Watson and Crick's 1953 *Nature* article, it would not be the exact

* I thank the Harvard University Press for permission to reproduce illustrations from *Sociobiology*. I would also like to thank Trevor Pinch and the Sociology seminar at the University of York, Steve Woolgar and the Department of Human Sciences seminar at Brunel University, and Robert Fox and the seminar on popularization at the Centre de recherches en histoire des sciences et des techniques, La Villette, Paris, for allowing me to discuss earlier versions of this paper. I would also like to thank E.O. Wilson and Robin Dunbar for their comments.

phrasing of the claim, it would be the picture of the double helix, with the phosphate chains like flat ribbons, the base pairs as rods between them. So when the *New Scientist* illustrates an article (Maynard Smith, 1985) with a picture of double helices in the sky, with lines coming down from them to the hands and feet of a man and of various animals, the editors can assume that most readers, whatever their discipline, can interpret this rather complex message as a statement about genetic control of animal behavior. The double helix stands for the physical basis of heredity the way two intersecting ovals enclosing a circle stand for atomic energy.

E.O. Wilson's *Sociobiology: The New Synthesis* can provide some examples of how pictures work when an author wants to appeal to readers outside the immediate discipline in which the conventions for reading these pictures are well established. It did not introduce any one image as powerful as that of the double helix or the Bohr atom; indeed Holton (1978) asserted that the lack of such a master image could limit the theory's popular appeal. But it is a useful place to start looking at pictures because it contains such a large and heterogenous collection of them, with samples of almost every kind of figure currently used in evolutionary biology. It links the kinds of illustrations one would expect to find in *Scientific American* – photographs and elegant drawings – to the graphs and visualizations of models one would expect in the *Journal of Theoretical Biology*.

Wilson's book is not typical of any genre of scientific publication. It looks like a textbook, but it is also a scholarly review and an overview that brings together material that was not in one place before; it is addressed to other biologists, but it is also meant for social scientists and the general public. In this analysis, I am going to imagine the reader as a non-biologist who comes to the book as a work of natural history, the way one might read Darwin's *Journal of the Voyage of the Beagle* or watch David Attemborough's *Life on Earth* on the BBC. Of course there are other possible, and quite distinct, horizons of expectation that could just as easily be brought to the book – those of the reader of Wynne-Edwards on group selection, or of W.D. Hamilton on kin selection, or of Desmond Morris's popular works on animal behavior, or of Wilson's own earlier work on insects, or those of the reader of criticisms of sociobiology by Steven Rose, R.C. Lewontin, or Stephen Jay Gould. The genealogy one gives the work – as a textbook-like survey of behavioral ecology, as a new synthesis of population biology and ethology, or as the latest in a line of hereditarian manifestoes – is a crucial step in one's reading of it. I construct this non-biologist reader of natural history as my guide because I want to think about how such a reader might interpret the book's pictures, how its images carry over from the specialized literature of biology into public discourse.

When *Sociobiology* was published in 1975, it was a best seller; it was greeted with both extravagant praise and angry criticism, and the controversy still continues today, with a number of books and scores of articles devoted to analysis of the book, its methods, and its context (references in Myers, forthcoming). Though reviewers, pro and con, often commented on how lavishly illustrated and visually imposing the book is, there has been little discussion of how this museum-like collection of images contributes to the book's remarkable persuasiveness. The book links various sorts of pictures together in a sort of montage that parallels but does not simply reflect the book's verbal text. I want to show how some of these links between different kinds of pictures and different kinds of authority are made. First, I will distinguish several categories of pictures, to show how they differ in their conventions and their rhetorical effects. Then, I want to show how these pictures relate to the text, and how they are juxtaposed with each other. Finally I want to apply this analysis to some other popularizations of sociobiological ideas. Though I will describe the pictures, readers may want to look at a copy of the book to see some of the pictures I do not reproduce.

2. From photographs to graphs

Chapter 26 of *Sociobiology*, which deals with "The Nonhuman Primates," begins with a figure representing an evolutionary model; it is headed "The Primates" and shows arrows moving from top to bottom connecting various headings of "The Prime Movers" to various "Adaptive Social Traits," sometimes leading along the way through a network of other adaptive traits such as

"Increased Manipulative Skills." The chapter ends with two large black and white photographs of "Temporary resting parties of chimpanzees in the Gombe Stream National Park..." (Figure 26-8). The problem of the chapter, if it is seen as a set of images, is to establish a link between them, between highly abstract diagrams and apparently realistic photos, between an evolutionary hypothesis and the observations of ethology. I am going to treat the photograph and the pathway diagram as two extremes of a set of categories of pictures (my Figure 1).[2]

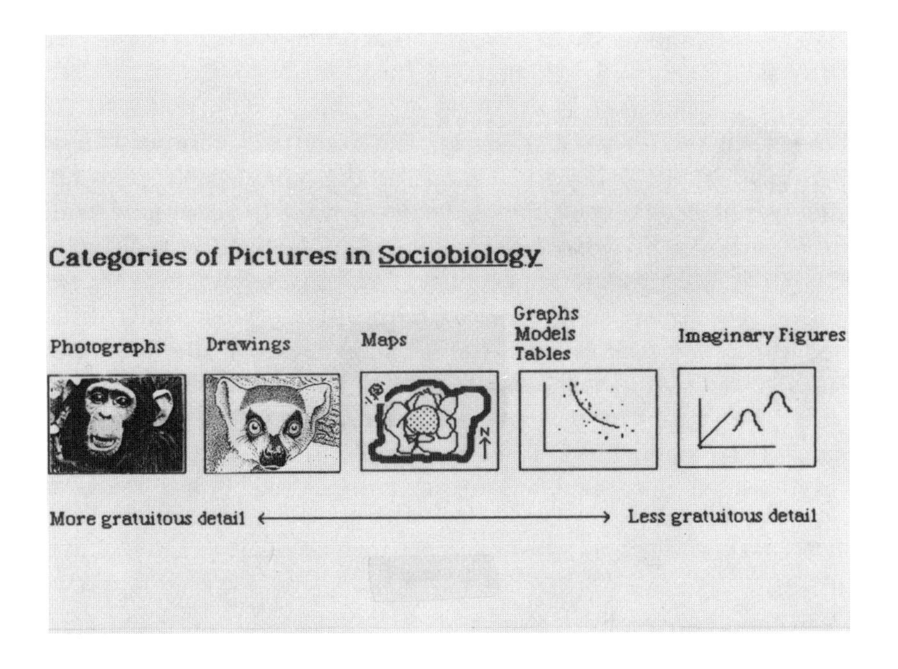

Figure 1. Categories of pictures in *Sociobiology*

One factor that distinguishes these categories is the presentation or elimination of gratuitous details. The squiggles and splotches that do not seem relevant to the claim the picture illustrates have their own significance, as part of the what makes the picture seem continuous with our own world. The elimination of these squiggles and splotches is part of the move from the particularity of one observation to the generality of a scientific claim.[3] I see a related change, from one category to the next, in the treatment of the background against which the object of the illustration is defined. At one end of this set of categories are *photographs*, which are full of gratuitous detail, and which present the background as a space continuous with our own. In *drawings* the artist gains the freedom to select and arrange details but loses the chaotic and arbitrary patterns of the photo; the background may recede in perspective or may be eliminated. *Maps* (of places or of bodies) are read as symbolic representations, rather than as images of the observed world. But in their backgrounds they still have some reference to the way we familiarly conceive of space, in the irregular outline of a waterhole, or the cutaway image of an ant's insides. In contrast, *Graphs, models*, and *tables* redefine space, wiping it clean of all irrelevant details and structuring it so that each mark has meaning only in relation to the presentation of the claim.

Photographs come with apparent self-evidence, because they are taken as mechanical reproductions of an image. As the film critic André Bazin (1945) has written, "Photography affects us like a phenomenon in nature, like a flower or a snowflake whose vegetable or earthly origins are an inseparable part of their beauty." But of course photographs are not part of nature, and are not entirely mechanical – the photographer selects the image and plays a part in defining it. To understand their effect, we must consider the way they are received, as well as the way they are produced. There has been enough written recently about photographs to make it clear that the interpretive process that turns dots into images is at least as complex as the optical and chemical processes that turn images into patterns of dots. Paradoxically, it is the limitations of photographs that make them such powerfully persuasive documents. The film theorist Rudolph Arnheim (1933) calls this a theory of "partial illusion"; the photographs cannot show time, space, or even images without our working on them. From the spotty grains of black and white, we reconstruct two-dimensional shapes and textures, and from the shapes and perspective we reconstruct volume and three-dimensional space, and from this reconstruction in space we reconstruct a moment in time.

First we must define some part of the photo as the image, usually setting it apart from the background. In Figure 26-8, the photos of the chimps (my Figure 2), our attention may go first to the pat-

Figure 2. A photograph. The caption reads: "*Figure 26-8* Temporary resting parties of chimpanzees in the Gombe Stream National Park. Left: Three adult males on the left (Worzel, Charlie, and Hugo) are accompanied by two adult females (Sophie, with a female infant, and Melissa). Right: in a second group, two infants play in the middle, one with a typical 'play face,' while a juvenile grooms an adult. (Photographs by Peter Marler and Richard Zigmond)."

tern in the middle because it is a light patch surrounded by dark, because it is in the middle, and because we look for the pattern of two dots above a curve signifying a face. The pattern at the top and the bottom is background because it is lighter, is incoherent, is at the edges, and, at top, is out of focus. The image is not always so easy to find. For instance, in a photo showing "slavery in ants" (Figure 17-7), Wilson must intrude arrows to enable us to pick out the relevant shapes of the master and slave ants. (One game show on British television sometimes challenges naturalists to pick out the animals in photos that apparently show a tangle of vegetation. But the audience must be shown highlighting, zooms to closeups, and enhanced outlines to make the image emerge.)

If we assume light from the top, and the shading conventions of realist, paintings, we can give this flat image volume; that is, we see this shape we have picked out as a face, and not, say, as a flat mask or cut-out.[4] Conventions of perspective direct us to read one chimp as being in front of another chimp and not, say, inside it. Just as important, conventions of perspective define our own place just outside the picture, just as the gaze of the chimp does. We place ourselves at a few feet from the chimp, and would be surprised if it had been taken with a telephoto lens from hundreds of feet away. This specific point of view is part of what gives a photo, or a realist painting, its immediacy and self-evidence. This point of view is also defined by the rectilinear edges of the photo. Just as we assume that a chimp at left was chopped off in framing or cropping, and is not really a whole image of half a chimp, we assume that the rest of the world goes on beyond the edges of this photo; this is one slice of a world that is continuous with our world. Perhaps the most powerful effect of a photo like this one is the way that, by freezing time, it suggests a narrative of events before and after – here, according to the caption, "resting," "playing" and "grooming." It is one particular moment, just as it is one particular place.

All these processes of interpretation lead us through a thicket of irrelevant detail. There is a great deal of detail here besides what the caption tells us to look for, "Two infants play in the middle, one with a typical 'play-face,' while a juvenile grooms an adult" (Wilson, 1975:544). Hundreds of dots go into the ear of an irrelevant chimp, or into a blurry stalk, or into the shadow under an eyebrow. All this detail carries no relevant information, but it does have a function, making the photo seem to be a document recording an unmediated perception of a particular piece of nature. The book also includes some photos that do not have all these signs of realism. In a photograph in which termites cluster around a bar of iron (Figure 14-2, 303) a photo taken from directly overhead, lit from behind, and greatly enlarged, the lack of a background, of shading, and of perspective in all suggest an experimental situation, rather than observation in the field.

Some critics, like Bazin, have begun their consideration of the photograph with the fact that it is a mechanical recording of the image: "All the arts are based on the presence of man, only photography derives an advantage from his absence" (13). But the rhetorical power of the photograph does not extend to other forms of mechanical recording of reality, such as spectrograms.[5] A spectrogram of a whale song (Figure 9-7, 221) carries no apparently self-evident message. This may be partly because the readers of popular natural history are less familiar with reading inscriptions that record frequencies of sound than they are with reading inscriptions that record the effect of light on an array of crystals. But there is also an assumption that photographs are organized like our ordinary sight, while spectrograms are organized like graphs. We only become aware of our processes of interpretation with photographs when we are looking for some trick in them. What both kinds of inscription share is a wealth of gratuitous detail. In the spectrogram, as in some photos, the relevant patterns must be picked out from the irrelevant details that are here labeled "artifact" and "dynamite."

Some of Wilson's critics commented on the number and lavish presentation of drawings in *Sociobiology*. They might seem to be out of place in a scientific book; they are not mechanical records and do not suggest that the artist ever saw just this, or even that the artist was in the place represented. But, like photographs, they use gratuitous detail and particularity to suggest immediate contact with reality. The most striking of these drawings are a series of two-page spreads by Sarah Landry, illustrating the social systems of various animals, each drawing based on one study cited in the caption; an example is the drawing of lemurs based on the research of Allison Jolly (Figure 26-3; my Figure 3). As in a photo-

Figure 3. Part of a drawing by Sarah Landry. The caption reads: "Figure 26-3 The encounter of two ring-tailed lemur troops at the Berenty Reserve in Madagascar. The habitat is a riverside gallery forest, dominated in the foreground by a large tamarind tree (*Tamarindus indica*). The arboreal troop on the left is stirring into activity after a noontime siesta. One male faces the observer with a threat stare, his antebrachial gland visible on the inside of the left forearm. A second male behind him has begun to move down the tree trunk in the direction of the other troop directly to his rear, two adults engage in mutual grooming, while other members stay clumped together in rest or in the early moments of arousal. The troop on the ground has begun its afternoon progression to a feeding site. Two adults at the left and front have spotted the group in the tree and are staring and barking in their direction. One of these, a male, draws its tail over the antebrachial glands in preparation for a hostile display. He is ready for a stink fight, during which the tail will be jerked back and forth to waft the scent toward the opponents. Well to the rear and in the center of this picture, two subordinate males of the 'Drone's Club' trail the second group. (Drawing by Sarah Landry; based on data from Alison Jolly, 1966 and personal communication.)"

graph, the frame, with vegetation and parts of animals continuing off it, and the perspective, fading to the dim small animals in the background, all suggest that the world is supposed to continue and this is just a metonymic piece of it. It is the strategy of the realistic novel. The observer is made present in the drawing of lemurs by the gaze of the male in front. Several of Landry's drawings have such inquisitive animals; one of the lions "stares at an unidentified object past the observer." This sounds odd; of course in a photograph the object would be unidentified, but in a drawing it, like the animals themselves, is imaginary.

A drawing uses these conventions of photographs, but it also allows manipulation in ways a photo does not. One handbook on popularization for scientists makes this distinction:

> If realism is important, the best choice may well be photographs. Otherwise, diagrams, which allow you to emphasize certain features and eliminate the rest, are often the best medium. (Gastel, 1983:13)

All of Landry's pictures arrange their components in a way a photographer cannot, to make them typical. They include in the same frame several representative activities, usually conflict between animals, recognition of the observer, sexual behavior, and the search for food, and these activities are usually read by the caption from left to right, and foreground (bottom of the page) to background. The drawings are deliberately impressionistic, pointillist, in style, and the layout makes them more like museum diorama than photos.[6]

Wilson emphasizes the representative quality of these drawings in his account of how they were made.

> I planned the panoramic views of animal societies for a special reason: unlike many other behavioral and biological phenomena it is impossible to represent more than a tiny slice of social life of any species with an ordinary photograph or sketch. Accordingly, I extracted information on the key members and some of the important behaviors from the best monographs and laid out the arrangement in a (crude!) pencil sketch. My work gave the two of us a lot of amusement. I took care to

> make the representation demographically correct and provided Landry with the original research materials. Landry took it from there, and added her own unusual gifts for scholarship and design. In many cases she visited zoos and watched motion pictures to sketch the live animals. She invented new arrangements and postures, including the striking one of the staring lemur. She researched the botanical literature and consulted botanists (both abundant at Harvard) to get as exact as possible the vegetation where the societies naturally live, and her drawings of individual plants are meticulous. (personal communication)

Wilson is careful to give full credit to his collaborator. He is also careful to stress the way that her careful research brought her into various kinds of contact with reality so that we will not look upon the drawings as inventions, as interventions of the artist or scientist.

Drawings can be classified on a continuum from the conventions of photographs to the conventions of diagrams. For instance, there is a Landry drawing of primitively eusocial bees, taken from an earlier Wilson book (20-7) that marks with a cutaway that we are seeing the unseen, the inside of a stem. And the reader would further qualify its realism because the stem is arranged in white space; the gratuitous detail in the background is eliminated, and there is no implication that the image extends beyond the frame. But even with this cutaway, the effect of realism is given by the thorns on the outside of the stem, the woody lines of the grain where the stem was cut, the shading of light from above, and the bee crawling out of the cut end. Abstracting further, the setting of an object can be eliminated entirely, and the pictures arranged like specimens mounted in a glass case; so, for instance, Figure 11-2 shows two rows of larvae, arranged with their heads to the top to display their mandibles, labelled with letters and a key. The shading can be eliminated so that only an outline is left, as in a drawing of a parasitic ant strangling a host queen (Figure 17-4). Here we are reminded that the use of black and white lines is governed by the conventions of prints, not by reality; we interpret this picture as meaning, not that a queen shown as white is white and an invader shown as black is black, but that the two animals are shaded differently so that we can distinguish them more easily. From such outline drawings focusing our attention on only one or two specific features, it is only a step to symbolic figures standing for the animals. In these symbols even more gratuitous detail is removed, so that a fish is an oval with fins, a bird is a profile silhouette of a beak, talons, and tail, and herds of rubber stamp elephants roam a landscape of wavy lines. Even so, such rubber stamp elephants suggest a reference in the picture to a visible reality that would not be suggested by an *E* with an arrow on a map.

A different sort of abstraction occurs in maps and diagrams. Martin Rudwick (1975) has discussed the problems of makers of early geological maps trying to represent three dimensions in two, using symbols or colors to represent the world below the surface. The problem in zoogeography is showing a series of events on a page: the distribution of animals, and the changes in this pattern, must be created out of a series of individual observations by the researcher over a period of time. The background is now defined by a two-dimensional grid, rather than by perspective, and by a few landmarks, rather than details of scenery. In many of these maps the space of a territory is defined by the a number of places an animal has visited or has marked over a period of time. The underlying set of conventions seems to be drawn from political maps, except that the boundaries are not drawn by fiat, like those of Merseyside or Loire-Atlantique, but are drawn through a series of conflicts, like the boundaries of Poland. In one map, for instance, the "travels of a coati band on Barro Colorado Island" are shown as a tangled skein of a line (Figure 12-1). The territory is then a boundary drawn around this tangle, and there is another theoretical construction, the "core area" further marked by shading within this boundary. Here all that matters is that there is a boundary; the exact shape of the territory is incidental, as is the path taken by the animals, and the gratuitous detail of its outline is effective just as testimony to its being a real, if unfamiliar, place.

The map framework can also be applied to bodies, in diagrams of deer glands (10-6) or of human hormones (11-5) or of the waggle dance of bees (8-1). As in maps, the object is reorganized as a space in two dimensions, and processes and entities can be marked as lines or arrows and regions within this grid. The outline of a body or organ serves as a terrain on which points are mapped. The theoretical entities are foregrounded, but there is still an underlying background representation of some real-world object. So, in a diagram of the waggle dance, the sun and the destination are represented by stylized symbols, the angles x and the ellipses tracing the movements of the bee are seen as mathematical abstractions, but they are marked against the background of woodgrain or of a honeycomb, as in representational drawings.

Pater said that all art aspires to the condition of music: in *Sociobiology*, all art aspires to the condition of a graph. While in a diagram the background, at least, refers to something we take as real, in a graph, the background is a theoretical space.[7] But graphs that record data retain some of the authority of automatic inscriptions by including gratuitous information. For instance, a graph of "frequency-dependent sexual selection," (Figure 15-2), with proportion of the population bearing a gene as the ordinate, and coefficient of mating success as the abcissa, includes a scattering of points and vertical bars that presumably show ranges. The clean line through them is given more, not less, authority by the fact that the points do not all fall exactly on the line. The points suggest accuracy and the likelihood of statistical variation in a finite number of trials, while the curve through them holds out the hope of a simple mathematical relation. Other types of graphs can be arranged in a continuum from those that look like messy printouts of recording devices to the clean and highly abstract correlations of ratios with ratios.

The figures that aspire to be graphs are the tables and the models. The tables are data points looking for curves. This is true even when the tables consist almost entirely of words, not quantities, so they may seem more a part of the text than of the pictures. They are not part of the text because the clusters of words are not to be read sequentially, left to right and then to the next line and so on, but are each to be read as a unit of information that is related to the other units in its row and column. A table showing the evolutionary grades of primates (Table 26-2) is part of a search for a pattern, a way of arranging these clusters of words so that the intersections of columns ("evolutionary grades") and rows ("ecology and behavior") will show relationships in a consistent development as one moves across the page. Wilson is frankly dissatisfied with this table, and with his own table of mammals, because such clear relationships have not yet emerged. He is happier with a table like 24-1, analyzing the social systems of ungulates into 'a relatively simple pattern that can be transformed with minor distortion onto a single axis, or "sociocline"' (479).

At the other extreme from graphs are the optimization models, which can be seen as curves looking for data points. As Wilson points out in an article (1977), the form of these graphs is borrowed from economics. The model for castes in insects is one of the key parts of the book (14-4, my Figure 4). The animals here are not treated as individuals at all, but are considered in terms of their collective weights.[8] There are many events behind this graph, such tasks as defending the colony, or foraging, or caring for the eggs. But there they are not even quantities; they are abstracted a further step and analyzed in ratios. A series of such graphs shows mathematically how shifts in environment or tasks can affect the caste structure. Behavior here is completely reduced to a set of laws. All the particularities of a photograph have been eliminated, and the space of the picture is now an entirely theoretical and quantitative space. But while some details are lost, some information is gained; a model like this can do things that a graph recording data cannot. The movement of a curve from one position to another gives a prediction of the ratios of the castes to each other; in this sense the world of the model replaces the world of the ant colony, just as the colony in the laboratory replaces that in the jungle.

The graphs and diagrams have an effect, even when the data in them are subjective, qualitative, or imaginary. So, for instance, Figure 2-2 (my Figure 5) shows what "The age-size frequency distributions of three kinds of animal societies" would look like in three dimensions, but the caption says, "These examples are based on the known general properties of real species, but their details are imaginary" (15). Some reviewers are annoyed by these graphs without actual data, taking them for lazy attempts at demonstration when they are, in effect, thought experiments on the page. That Wilson included them shows how important it is for him to visualize even the most theoretical statements. To

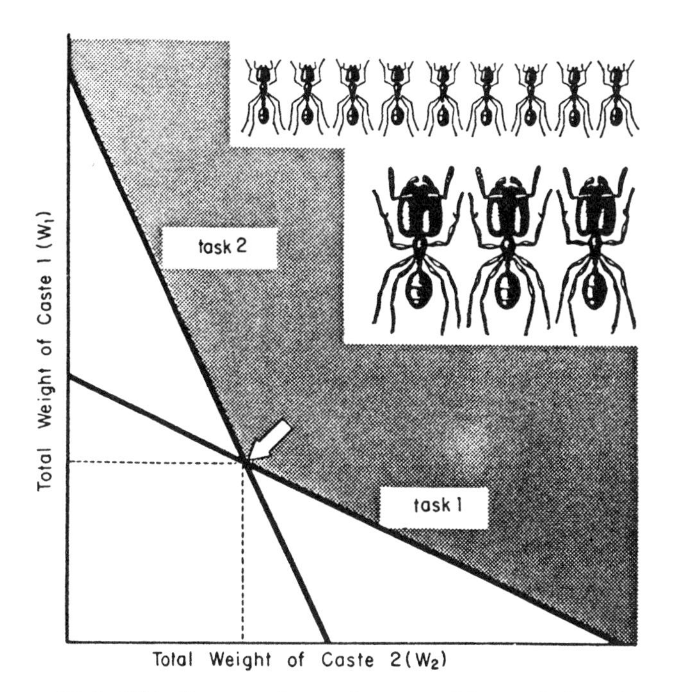

Figure 14-4 This diagram shows the general form of the solution to the optimal-mix problem in evolution. In this simplest possible case, two kinds of contingencies ("tasks") are dealt with by two castes. The optimal mix for the colony, measured in terms of the respective total weights of all the individuals in each caste, is given by the intersection of the two curves. Contingency curve 1, labeled "task 1," gives the combination of weights (W_1 and W_2) of the two castes required to hold losses in queen production to the threshold level due to contingencies of type 1; contingency curve 2, labeled "task 2," gives the combination with reference to contingencies of type 2. The intersection of the two contingency curves determines the minimum value of $W_1 + W_2$ that can hold the losses due to both kinds of contingencies to the threshold level. The basic model can now be modified to make predictions about the effects on the evolution of caste ratios of various kinds of environmental changes. (From Wilson, 1968a.)

Figure 4. A model

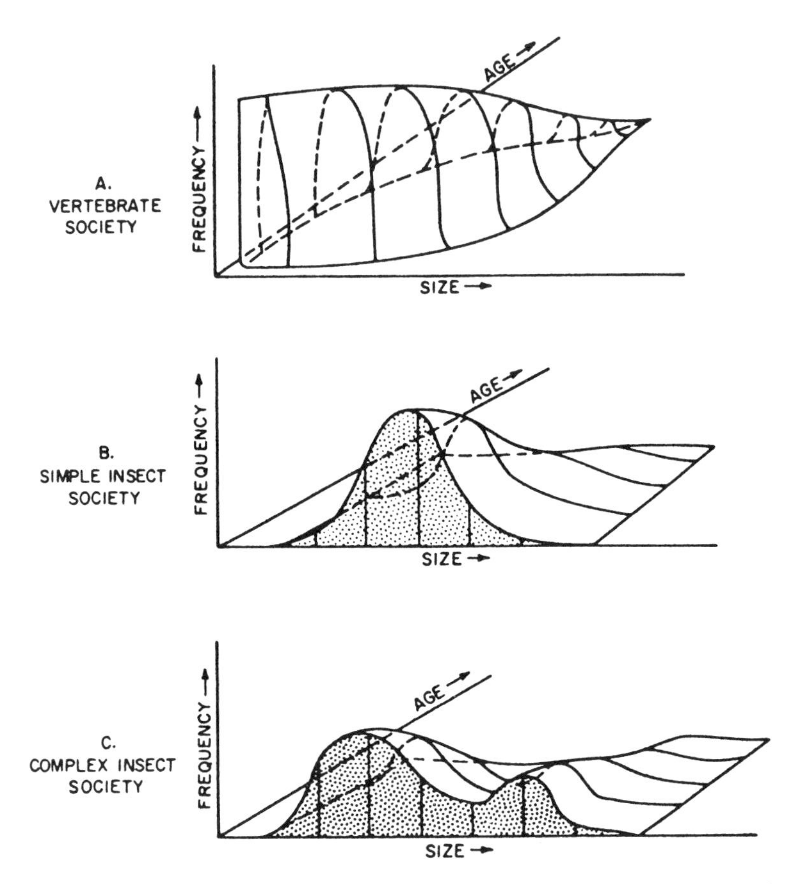

Figure 2-2 The age-size frequency distributions of three kinds of animal societies. These examples are based on the known general properties of real species but their details are imaginary. *A*: The distribution of the "vertebrate society" is nonadaptive at the group level and therefore is essentially the same as that found in local populations of otherwise similar, nonsocial species. In this particular case the individuals are shown to be growing continuously throughout their lives, and mortality rates change only slightly with age. *B*: The "simple insect society" may be subject to selection at the group level, but its age-size distribution does not yet show the effect and is therefore still close to the distribution of an otherwise similar but nonsocial population. The age shown is that of the imago, or adult instar, during which most or all of the labor is performed for the colony; and no further increase in size occurs. *C*: The "complex insect society" has a strongly adaptive demography reflected in its complex age-size curve: there are two distinct size classes, and the larger is longer lived.

Figure 5. An "imaginary figure"

do this, he draws on the same conventions of maps, grids, and curves he uses in showing data, but uses them here to create purely theoretical spaces. Of course Wilson was not the first to use such representations of evolutionary abstractions; Darwin's evolutionary tree is also a visualization of a model, and Wilson uses C.H. Waddington's "adaptive landscape" in 2-6. What intrigues me about this latter kind of illustration is that it seems to come full circle; the highly abstract concept of a surface area corresponding to genotypes and a vertical measure of a quality, fitness, comes out looking like a representation of a perspective on a real world.

Wilson's usual way of visualizing a theory is a sort of flowchart, like that showing the evolution of primate social behavior (Figure 26-1; my Figure 6) in which a series of causal connections is represented as a movement on the page. The background is blank, instead of being a grid, and there is not even a definite convention about the representation of time, which is generally left to right in graphs. Here, time is sometimes read from left to right (22-2), sometimes bottom to top (12-14), sometimes top to bottom (2-8), often both directions at once (15-11). All that matters are the connections of the arrows and the clusters of words to which their heads point; we can read them without an explicit caption just as we can read maps of Turner's first tour of Yorkshire.[9] One difference between the pathway diagrams and such tour maps is that the arrows in these theoretical narratives are causes, not events happening to one actor. So they are the furthest step of Wilson's argument: a table suggests a correlation, a graph shows it in quantitative terms, and the pathways make it a step in a series.

I have arranged these categories in terms of realism and abstraction. From one category to the next, gratuitous details are eliminated, and the background is transformed from a space continuous with that of our everyday worlds to grid. The elements of narrative events, time, space, and subject, are transformed. In photographs, the image becomes a narrative. In pathway diagrams and adaptive landscapes, the narrative of evolution becomes a two-dimensional image.[10]

The distribution of pictures from these categories through the

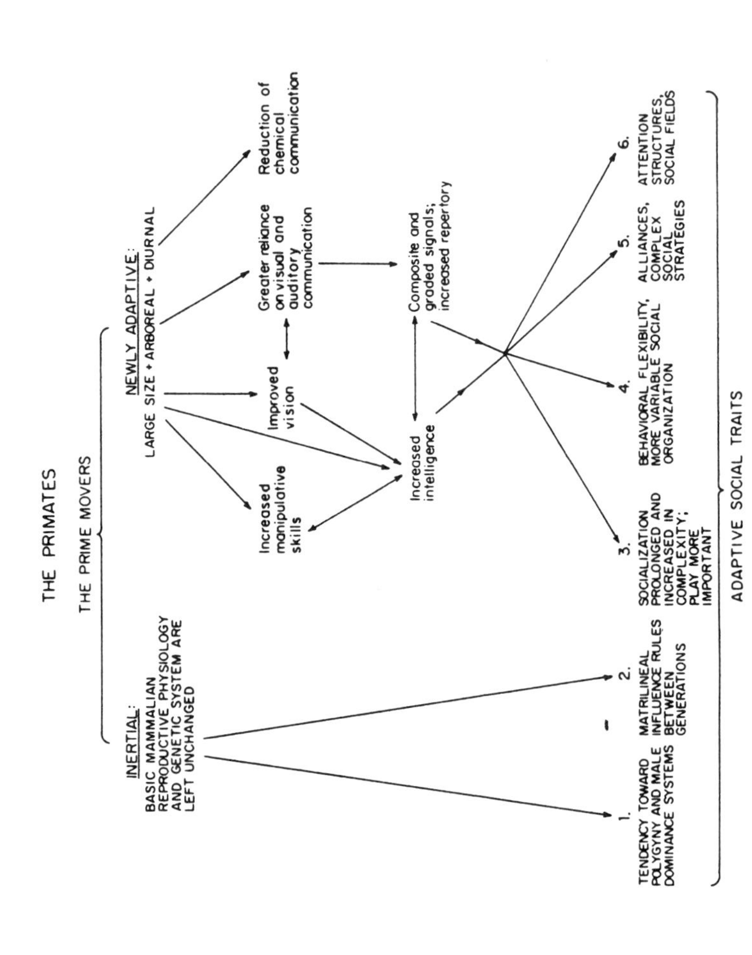

Figure 26-1 The distinctive social traits of the higher primates are viewed as the outcome of conservative mammalian qualities ("inertial" forces) and adaptation to arboreal life. Even phyletic lines that are now terrestrial have retained the evolutionary advances made by their arboreal ancestors.

Figure 6. A pathway diagram

text can indicate the kind of argument going on. As we might expect, the theoretical opening chapters of the book are illustrated mainly with diagrams and graphs; for instance, Chapter 4, "The Relevant Principles of Population Biology," has 22 figures and two tables, but no photographs or drawings at all. Nearly all these pictures demonstrate some part of the argument, so, in Figure 4-1, a graph shows that a computer simulation of genetic drift does rapidly lead to fixation of one allele and loss of another. On the other hand, the pictures in the last nine chapters that survey the animal kingdom consist nearly entirely of tables that survey the part of the taxonomic system being discussed in the chapter, and of photos and drawings that simply illustrate statements in the text. It would seem that the more abstract figures go with the more abstract and difficult chapters. But Chapter 3, "The Prime Movers of Social Evolution," which one might expect to be as theoretical as the chapter on population bilogy, has only two of fourteen figures that don't have some photo, drawing, or symbolic representation as part. We need to look, not just at individual pictures, but also at how the picture fits into the text, and how one picture affects the reading of those before and after it.

3. Words and pictures

Of course the reader interpreting these pictures does not just rely on conventions associated with photographs, drawings, maps, and graphs. Every illustration in *Sociobiology* is associated with two passages in the text: the caption, which tells us what we are seeing, and the sentence in the text marked with the reference to the figure, which tells us how it fits into the verbal narrative. This is the caption for two pictures of chimpanzees, one of which is the photograph with which I began.

> *Figure 26-8* Temporary resting parties of chimpanzees in the Gombe Stream National Park. Left, three adult males on the left (Worzel, Charlie, and Hugo) are accompanied by two adult females (Sophie, with a female infant, and Melissa). Right: in a second group, two infants play in the middle, one with a typical "play face" while a juvenile grooms an adult. (Photographs by Peter Marler and Richard Zigmond)

Usually the first part of a caption, or in some cases the only part, is a long noun phrase, not a sentence, that functions as a sort of title. The caption leads us through the thicket of gratuitous details to the shapes that are supposed to be meaningful. So we look, for instance, at the face of the small chimp in the center, rather than at the chimp covering its mouth at left. Captions tell us what happened before, what will happen after, what usually happens; so, for instance, this caption tells us that both groups are only temporarily resting, and that the figure in the centre of the photo on the right is an infant that is playing. The caption helps link each picture to the next level of abstraction: here the caption in effect emphasizes and abstracts the "play face," in effect making a diagram of the photo. Other captions can make a drawing into a diagram by pointing to one specific movement, and bringing out the causal connections in the graphs and diagrams.[11] The caption also gives the authority for the image, so that all the photo-captions end with photographers' credits, and all Sarah Landry's panoramas end with a textual source, like the caption to the lemur drawing, which ends '(Drawing by Sarah Landry; based on data from Alison Jolly, 1966 and personal communication).'

In *Sociobiology*, as in most scientific books and articles, there is always another reference to the illustration in the text, as well as the caption. So, regardless of the demands of layout that may put an illustration on one page or the next, each illustration is formally inserted at an exact point in the text. It can be inserted there either to illustrate a point made in the text, or to demonstrate and extend the text. For instance, the reference in the text to the lemur illustration mentions this whole elaborate panorama only as a way of pointing to the position of an underarm gland: 'Brachial glands, which occur high on the male's chest, and conspicuous antebrachial organs on the forearms produce odorous substances (see Figure 26-3).' Here it is not even necessary to have the picture; it is a supplement to the text, adding details or specificity, or illustration of it. In contrast, the references to many of the graphs are not complete without a study of the graphs themselves. For instance, Wilson refers to his ergonomic model in the sentence, 'The general form of the solution to the optimal mix problem is given in Figure 14-4.' One way of distinguishing these two uses of pictures is in the form of the reference; a parenthetical figure number often implies an illustration, while demonstrations usually include the number in the structure of the sentence. So, syntactically, the demonstrations are part of the linear flow of the text, while the illustrations are parallel to it but set off from it.

Some textual explication is necessary even for pictures that seem to carry a self-evident meaning. Two photos in Figure 3-13 demonstrate that fire ant workers respond to evaporated trail substances; the difference between the photos before and after trail substance is blown over the ants is to be taken as proof of the claim. In the top photo, there is a dish on the left with black specks in it, and on the right a watch in the background and in front of it a block of wood with a glass rod on it. The bottom is the same except that the hands of the watch have moved and there are black specks to the right of the dish. The insertion of the watch, an otherwise odd element, adds to the effect of self-evidence by giving apparently undeniable evidence that the two photos are separated only by "a short time." But to see what claim being proved, one needs to be directed by the caption, and informed about what these black specks are and what was done between the two pictures.

> *Figure 3-13* The response of fire ant workers to evaporated trail substance. Above: before the start of the experiment, air is being drawn into the nest (by suction tubing inserted to the left) from the direction of the still untreated glass rod. Below: within a short time after the glass rod has been dipped into Dufour's gland concentrate and replaced, a large fraction of the worker force leaves the nest and moves in the direction of the rod. (From Wilson, 1962a:56)

Without these instructions from the caption, one might see the pair of photos as demonstrating that iron fillings are attracted by a bar magnet, or that ants are attracted to stopwatches.

4. Pictures into stories

In a book with so many pictures, or in a lecture, a popular article, or a television show, part of our interpretation of each picture depends on how it relates to those we have already seen. One picture can be related to others when it is part of a pair or a series, or when it has the same form or subject matter as another picture. The photos of the experiment with the stopwatch are meaningful only as a pair, and the series of graphs of the ergonomic model of castes makes sense only as a series. One set of six drawings (Figure 8-3) show, from left to right, three stages of the aggressive display of a rhesus monkey, and below that, three stages of the aggressive display of a green heron, each with arrows showing the movements of the head or the tail feathers. The individual drawings are as realistic as any in the book, but they are placed here in an abstract space, in which time passes from left to right, so each row forms a narrative, while comparison of the top and bottom rows suggests similarities in behaviors across species. The abstraction is a crucial move, for some comparative psychologists would deny that this kind of comparison can be meaningful. Any series of pictures makes a theoretical statement by bringing out selected features from the innumerable details of the pictures, and putting them in narrative order.

Juxtapositions between different sorts of figures that are not obviously part of the same sequence can also be significant. The drawing and photos are often linked to graphs or pathways, when one type of illustration immediately follows another or has the same stated topic as another, or has a similar form. Many of the graphs include some symbolic representation of the animal involved; even Wilson's highly abstract ergonomic models of ant castes have symbolic ants in the upper right corners. The figure then brings together the more representational and more abstract pictures, the illustrations and the demonstrations. For instance, Figure 2-4, shows, at bottom, bar graphs of the percentages of time various species of monkeys devote to such behaviors as grooming and observing, while at top it has line drawings of monkeys performing these behaviors.

The famous diagram illustrating "altruism," "selfishness" and "spite" in their sociobiological senses (5-9; my Figure 7) and the graph that follows it show how two figures on the same stated topic can support each other. In the diagram the concepts are reduced to black and white circles with circular happy faces or

Figure 5-9 The basic conditions required for the evolution of altruism, selfishness, and spite by means of kin selection. The family has been reduced to an individual and his brother; the fraction of genes in the brother shared by common descent ($r = \frac{1}{2}$) is indicated by the shaded half of the body. A requisite of the environment (food, shelter, access to mate, and so on) is indicated by a vessel, and harmful behavior to another by an axe. *Altruism:* the altruist diminishes his own genetic fitness but raises his brother's fitness to the extent that the shared genes are actually increased in the next generation. *Selfishness:* the selfish individual reduces his brother's fitness but enlarges his own to an extent that more than compensates. *Spite:* the spiteful individual lowers the fitness of an unrelated competitor (the unshaded figure) while reducing that of his own or at least not improving it; however, the act increases the fitness of the brother to a degree that more than compensates.

Figure 7. The diagram of altruism, selfishness and spite

unhappy faces. They carry either a jug, representing some resource, or an axe, representing harm. The oddity is that the figures are so abstract, but the illustration still appeals to our common sense definitions of the words and categorization of everyday objects. This figure of apparently naive simplicity is followed by a graph (5-10) that attempts to quantify the level at which an altruist gene would become fixed. It is an example of the most abstract kind of graph, in which neither axis is a quantity like time or number of organisms; instead, both are proportions of 1, and the equlibrium point is a critical point on one of these proportions in relation to the other. The two illustrations play off each other, one pointing in the direction of daily experience, the other pointing in the direction of mathematical abstraction.

The chapter on territory shows the alternation of more realistic and more abstract figures, in which photographs and maps lend particularity to the graphs and pathway diagrams. For instance, Figure 12-8 shows "territories of the dunlin, a species of sandpiper that breeds in Alaska," on a conventional map. The empirical nature of the figure, the fact that these maps refer to real places, Kolomak and Barrow, and to real observations, in 1966 and 1962, is suggested by the way the lines of territories cross the edges of a grid, extending beyond the frame. These maps are related by the narrative of the chapter to an even more empirical figure, a photo showing fish in their own territories (12-9), taken from overhead so that the boundaries of heaped-up sand that the fish have made show in the pattern of light and shadow as an array of hexagons. This photo is crucial to the iconography of the chapter, for it shows a map of territories that is an inscription of the animals themselves, suggesting that "territory" is not a theoretical concept created by biologists in interpreting their observations, but is a fact that can be directly observed, a disk of sand with walls around it created by the fish.

These and the other territory diagrams in this chapter help us understand one of the most controversial illustrations in the book, the territory diagram of disciplines in the first chapter (1-2). In that diagram, Wilson shows, "A subjective conception of the relative number of ideas in various disciplines in and adjacent to behavioral biology to the present time and as it might be in the future." In this conception, sociobiology expands at the expense of neighboring disciplines; the picture is a bit baffling to nonbiologists beginning the book because it is hard to see what these rounded, bread-dough-like forms on two different planes are supposed to represent. It becomes clearer when we come later to a figure like 12-15, "Territorial exclusion in two species of blackbird." The parallel in visual conventions suggests that scientific specialties (in 1-2), like birds (in 12-15), compete for limited resources, and that what one gains must be taken from another – an assumption that is defensible, but is by no means universally accepted.

5. Pictures as icons

I have argued, in discussing the controversy over *Sociobiology*, that quotation is always quotation out of context (Myers, 1986). In the same way, the re-use of an illustration, even if the source is an earlier work of Wilson's, changes the meaning of that illustration. Sarah Landry's illustrations to Wilson's own *Insect Societies*, reproduced in *Sociobiology*, now illustrate the nature of social systems while before they just illustrated observations of insects. When Wilson reproduces a detail from Sarah Landry's *Sociobiology* illustration of the wild dogs in his own article in the *New York Times Magazine*, 'Human Decency is Animal,' (1973b), he alters the context and the caption. In the book, it is introduced with the phrase, 'The "super beasts of prey" and most highly social canids: a pack of wild dogs on the Serengeti Plains of Tanzania ...' (510). In the article, it is introduced with the phrase, 'Responsibility: In Tanzania, a wild dog, home from the hunt ...'. The change of context turns the picture from an illustration of one type of nomadic carnivore society, built around maternal care, into a general moral fable about the need for community if the species is to survive.

Similarly, Wilson's illustrations are appropriated in reviews and comments on the book. These uses can be allusive or ironic, but can never be straight. So, for instance, the drawings by Sarah Landry selected for reproduction in reviews in *Nature* (Wynne-Edwards, 1976), *La Recherche* (Hopkins, 1977), and the *New York Times Book Review* (Pfeiffer, 1975), and the first news report in the *New York Times* (Rensberger, 1975), all carry a meaning besides what they were illustrating in the book. The illustration used on page 1 of the *New York Times*, showing porpoises lifting a wounded comrade, connects *Sociobiology* to popular books (which Wilson criticizes severely) that suggest porpoises have a culture parallel to our own. The gorillas in *Nature* (and other gorillas in the *New Scientist*) make the connection between sociobiology and the Descent of Man. The lemur in the *Book Review*, and in *La Recherche*, as we have seen, attracts attention with its goggle-eyed animal staring out at us, and with its appearance of a cartoon figure. So all these illustrations stress the relation to man that is the most controversial aspect of *Sociobiology*.

Other photos and drawings seem to be quoted in reviews and in commentary to represent the public image of the field of sociobiology. The most obvious way of doing this, as with any science, is to provide a photograph of the scientist; on the front page of the *Times*, next to the porpoises, he is seen as a thin, middle-aged man in horn-rimmed glasses with a half smile, the epitome of the academic. The jacket photo on the back flap of *Sociobiology* (Figure 8) portrays him with arms folded, in front

Figure 8. The jacket photograph of E.O. Wilson. (Photo by Chris Morrow)

of a display case containing, I think, a gibbon and an orangutan. A *New Scientist* article has a picture of Wilson in front of a (presumably stuffed) gorilla. When the *New Scientist* editors fished this picture out of their files for a retrospective, they mischievously labelled it "E.O. Wilson, foreground." Now Wilson studies ants, but he is never shown with ants, and he has little to say about orangutans, or about any apes but the chimpanzees. But the pose seems appropriate because of the nineteenth-century iconography of Darwin confronting his "ancestor" the ape (Figure 9).

Figure 9. A Victorian cartoon of Darwin and the ape

Quoted illustrations can be used ironically, even in apparently straight quotations. So, for instance, Joe Crocker's *Radical Science Journal* critique of sociobiology (1983) reproduces the diagram and caption defining altruism, selfishness, and spite. In *Sociobiology*, this was meant to illustrate and distinguish the three sorts of behavior with an abstract example, but in its new context it is meant to demonstrate the absurd anthropomorphism of sociobiology, the cartoon-like happy faces suggesting a naive simplicity. The same illustration is used without apparent irony in *La Recherche*; but here the words are not quoted directly (or even translated literally, though the translation is fairly close). Since the intention in *La Recherche* is not ironic, the exact words of the caption are not essential. The figure illustrates a concept, instead of demonstrating a flaw.

Some later references to *Sociobiology* use caricatures rather than direct quotations. *New Scientist* illustrates an attack on *Sociobiology* by the Science as Ideology Group with a caricature of ants as Nazis. This alludes to Wilson's specialty, and to the authors' argument that he sees reactionary tendencies in animal societies by projecting human societies onto them. Two recent retrospective articles also make a comment on *Sociobiology* in their cartoon illustrations. I have mentioned the *New Scientist* illustration of double helices dangling by marionette strings, a lizard, a man in a business suit, a bird, an ant, and an elephant. This is used even though Maynard Smith's article is cautious about the applications of sociobiology to humans. Critics of sociobiology have referred to a similar cover picture on the issue of *Time* that reported on sociobiology. (Similarly, the cover of a French popularization of sociobiology [Christen, 1979] shows computer graphics contour represention of a man, woman, and child, suggesting that sociobiology leads to technical control and standardization of human life.) The cartoon in another retrospective article, by John Krebs, shows two fish drawing their strategies on blackboards like football captains. The joke is in the juxtaposition of the abstract representation of strategies in game-theory approaches to the evolution of behavior with the realistic image of fish acting as conscious strategists in characteristically human roles.

The article by Joe Crocker in the *Radical Science Journal*, to which I have already referred, has a cartoon of three monkeys wearing hats that, in England, are taken as upper class, middle class, and working class symbols. This is a parody of Figure 13-3 in *Sociobiology*, a photograph showing rank order of rhesus monkeys, as indicated by direction of grooming. Such parodies make us aware of the conventional nature of interpretations that are otherwise accepted as natural. Here the parody plays with the social reading of the hats, and the ethological reading of the grooming, and with the representation of dominance, which can be observed only in terms of social behavior and interactions, as if it were a hierarchy fixed in space. The cartoon, like the article, suggests that ethological readings have an origin in taken-for-granted assumptions about class in human society.

6. Pictures tell a story

In many ways *Sociobiology*, as a lavishly illustrated survey for biologists that is also aimed at the general academic reader, is an anomaly in a world of texts divided between the popular and the professional. But even if there are no exactly comparable texts, its anthology of illustrations makes it a useful starting point in a discussion of illustrations in popularizations of evolutionary ideas. While more popular presentations do not use the battery of graphs and tables found in *Sociobiology*, they do select their images from a range of categories, each of which carries its own conventions of interpretation, and they do juxtapose several kinds of images. In many popularizations, as in *Sociobiology*, the juxtaposition links pictures that have the authority of our everyday experience of the world to pictures that carry the authority of science.

In future research on the uses of illustrations on the popularization of science, it might be useful to compare this analysis to that of popularizations in other media or other periods. As the recurrence of the image of Darwin and the apes shows, the iconography of popular science is remarkably persistent, so we might expect to find some of the same images in the famous nineteenth-century popular science lectures and in popularizations of sociobiology. On the other hand, new technologies of reproduction of images change the sorts of images that can be used, so *Sociobiology* is a different sort of book from any in the nineteenth century, and it is different from presentations of the same ideas on television.

Nineteenth-century lectures, like *Sociobiology*, often use images of an apparently real world to lead to theoretical narratives, so, for instance, T.H. Huxley starts his most famous lecture with a piece of chalk as a prop, and Louis Agassiz starts one of his *Geological Sketches* with a description of a particular landscape. Both take this solid point of departure for their very different visions of the geological past. The formats of more experimental lectures, like those by John Tyndall on *Heat as a Mode of Motion*, are often built around the problems of making the large audience see the phenomenon, often through projected or enlarged displays; in the published versions all the paraphernalia survive only in the crowded engravings of equipment in the plates. A comparison of Michael Faraday's 1861 lectures on *The Chemical History of a Candle* with this year's televised Royal Institution Christmas Lectures might make one think that technological changes were relatively unimportant in the presentation of popular science, for Lewis Wolpert used the same sort of play with objects and models that his predecessor used. Yet it could be argued that, even in such a conservative genre as the Christmas Lectures, everything has changed. Television, showing a representation of a lecture rather than a lecture, makes the lecturer another of the objects to be watched, like the candles or tubes full of ping-pong balls.

Richard Dawkins' recent program *The Blind Watchmaker*, part of the BBC's Horizon series, is comparable to *Sociobiology* in presenting an argument for adaptation that combines pictures of animals with attempts to visualize an abstract model, so it may suggest some of the possibilities and difficulties of extending this sort of analysis to television.[12] One sequence, for instance, alternates images of animal eyes with images of the scientist and of designed objects. As Dawkins summarizes Paley's argument that the intricacies of the eye prove design in nature and the existence of an intelligent creator, we see him inside a gothic window at Oxford, looking in an antique microscope; in effect he is the modern scientist playing for a moment the eighteenth-century natural theologian. Then we see closeups of a fish's eye, and of a moth's eye, and then several shots of Dawkins holding and using a computer disk (the disk, in a joke, has the first verses of Genesis on it). An electron micrograph of the moth's eye is followed by electron micrographs of the computer disk, to show how the surface of the disk imitates the surface of the moth's eye. The use of gratuitous details is like that in *Sociobiology*, but on television the details that lend a sense of realism to these images can be on the soundtrack as well as in the pictures (and the sounds were presumably recorded after the pictures); the soft splashes of the fish, the beating of the moth's wings, even the noise of the holes being punched in the computer disks. As with the photos in *Sociobiology*, there are too many details for viewers to perceive them all. But the voice and presence of Dawkins guides us like a caption to what we are supposed to look at. The techniques of montage also have some of the functions of the captions in a text; the alternation of medium shots, close-ups, and microscopic images links the visible eye and the visible disk, apparently so little alike, through their similar images under the electron microscope. These microscopic images are linked through a peculiarly cinematic technique of panning in the same direction across their surfaces.

But television is not limited to the presentation of natural history photographs, to the first of the categories of pictures I described, as one might think. Later in the program, Dawkins goes on to present a fairly abstract model of Darwinian evolution by natural selection, through a computer simulation of "biomorphs" controlled by "genes" that vary within given limits and are selected by the operators' preferences.[13] At the end of the program, Dawkins' "selfish gene" is illustrated with a close-up view through an animated DNA double helix spinning in space like a starship in a science fiction movie, which then draws back and wraps itself around histones and coils into a chromosome, which recedes into a spot on the screen, which turns into a fish and then into a dinosaur. This kind of fluid juxtaposition is possible only with computer animation, but the idea is the same as that of the cartoons using the double helix for Maynard Smith's *New Scientist* article. Here the world of visible organisms (if one can include dinosaurs in that world) and the unseen structures of information proposed in the theoretical model are seamlessly linked, literally in one line.

Sociobiology does not have color, pans, sound effects, voice-over, or computer animation. But it achieves some of the same effects with the juxtaposition of the sober black and white of its photographs, the impressionistic dots of its drawings, and the severity of its graphs. I have paralleled the effect of this juxtaposition to cinematic montage, and this parallel should remind us that the effect of realism does not depend on the complete reproduction of the world, but on the viewer's perception of the narrative and perceptual order. Arnheim comments that one can edit together images only because they are not recreations of real life. "If film photographs gave a very strong spatial impression, montage would be impossible. It is the partial unreality of the film picture that makes it possible" (28). In the same way, the partial unreality of the images in *Sociobiology* (or in any printed popularization), requires us to reconstruct the space within them, and allows us to link photos to maps or drawings to graphs and to produce stories out of pictures. Paradoxically, for popular readers at least, the work we must do to put all those pictures together is what makes the story they tell seem so powerful.[14]

Notes

1. Martin Rudwick, in his pioneering study of the visual language of nineteenth-century geology (1976), attributes this oversight to the lack of a tradition of using visual communication in historical research – history is an overwhelmingly verbal discipline, as are sociology and linguistics.
2. François Bastide (1985) has categorized scientific illustrations in terms of the increasing intervention of the authors in the image and the increasing complexity of the information contained. In the same volume Mike Lynch has related electronmicrographs to graphs through different kinds of images in a process he calls *mathematicization*. I am going to present similar categories, but will discuss them in terms of some formal features of the images and their rhetorical effects on my imagined reader.
3. This would seem to be the opposite of Bastide's reading, in which the more abstract presentations can contain more semiotic dimensions. I think we are describing the same categories, but she is talking about what is added, and I am talking about what is taken away.
4. Bastide (1985) discusses the ambiguity of this convention.

5. A mechanical record of the head-bobbing patterns of a lizard (8-11), or even a lifelike electron micrograph (14-3), are other records that suggest reality because of the apparent exclusion of manipulation (on electron-micrographs, see Lynch, 1985b).
6. But oddly, even when there are these signs to the contrary, a reader takes these texts as showing what really is. Two of Landry's drawings are clearly labelled "speculative" in their captions; one shows dinosaurs roaming the plains and the other shows a group of early men (and no women) fighting off rival carnivores and tucking into a dead mammoth (27-5, 570-1). Some reviewers object to the portrayal of early man in terms of hunting (see, for instance, Pfeiffer, 1975). Despite the clear disclaimer in the caption, the effect of the drawing is to suggest that men were evidently carnivores. This effect is supported by the caption that gives warrants for everything else in the drawing: the plants, the volcanoes in the background, the prey, the competitors, the tools in their hands, even the shape of the saber tooth cats. But most of all it is supported by the conventions of representation in drawing. As Gilbert and Mulkay (1984) have pointed out, it is very difficult to mark a realistic drawing as speculative.
7. Laura Tilling (1975) traces early graphs to the inscriptions of recording devices like thermometers and barometers. I am arguing that we read machine inscriptions and graphs differently, even though they look the same. Even a graph that takes time as its ordinate, like a recording machine printout, treats that time as a quantity rather than as a continuing process.
8. Michael Lynch (1985a), in his studies of a neurobiology laboratory, describes the transformation of "animals" from individual organisms into dated and labelled experimental units.
9. The arrows themselves may be altered to carry various kinds of significance. In a flowchart of grasshopper behavior (9-6), different thicknesses of arrows indicate the number of observations of each transition between states. In one diagram, Figure 10-3, the original author, van Hoof, tried to suggest the indefiniteness of some evolutionary connections with shady bands that look like swarms of bees. Wilson sometimes uses dotted arrows, or two different arrows, to suggest alternate interpretations. The problem here is finding the visual equivalent of hypothesis, as in the drawing mentioned in note 5.
10. Françoise Bastide (1985:144), makes a similar point about the relation between graphs and photographs.
11. Some of the maps and graphs just have the first long noun phrase of the caption, as a sort of title, but usually they go on in an independent clause to explain just how we are supposed to interpret the graph. The only illustrations that just get a title phrase are some pathway diagrams (10-3) and tables. This lends support to the idea that captions are necessary to enable the readers to find their way through the thicket of details; only in the final stage of abstraction or the first stage of sort-

ing out data are the intended meanings clear enough.

13. If some confused viewers take this to mean that Dawkins does accept Paley's argument from design, but believes that God has changed his profession from watch-making to computer programming, that only shows that a sequence of images can always carry meanings beyond those intended.

References

Arnheim, R. (1933). Film and reality, trans. in *The Film as Art* (1957). Berkeley: University of California Press.

Bastide, F. (1985). Iconographie des textes scientifiques: Principes d'analyse. In B. Latour and J. de Noblet (Eds.), *Les "vues" de l'esprit*. Special issue of *Culture Technique* 14 (June).

Bazin, A. (1945). The ontology of the photographic image, trans. in *What is Cinema* (1967). Berkeley: University of California Press.

Christen, Y. (1979). *L'heure de la sociobiologie*. Paris: Albin Michel.

Crocker, J. (1983). Sociobiology: The capitalist synthesis. *Radical Science Journal*.

Gastel, B. (1983). *Presenting science to the public*. Philadelphia: ISI Press.

Gilbert, G.N., and Mulkay, M. (1984). *Opening Pandora's box: A sociological analysis of scientists' discourse*. Cambridge: Cambridge University Press.

Holton, G. (1978). The new synthesis? In M. Gregory, A. Silvers and D. Sutch (Eds.), *Sociobiology and Human Nature*, 75–97. San Francisco: Jodssey-Bass.

Hopkins, P.O. (1977). La sociobiologie. *La Recherche* 8 (75):134–142.

Jacobi, D. (1985a). La Visualisation des concepts dans la vulgarisation scientifique. In B. Latour and J. de Noblet (Eds.), *Les "vues" de l'esprit*. Special issue of *Culture Technique* 14 (June).

Jacobi, D. (1985b). References iconiques et modèles analogiques dans des discours de vulgarisation scientifique. *Information sur les Sciences Sociales/Social Science Information* 24:847–867.

Krebs, J. (1985). Sociobiology ten years on. *New Scientist* (3 October): 40–43.

Latour, B. (1985). Les "vues" de l'esprit, in B. Latour and J. de Noblet, trans. Visualization and cognition. In H. Kuclick (Ed.), *Knowledge and society: Studies in the sociology of culture, past and present* 6:1–40. Greenwich, CT: Jai Press.

Latour, B., and de Noblet, J., Eds. (1985). *Les "vues" de l'esprit*. Special issue of *Culture Technique* 14 (June).

Lynch, M. (1985a). Discipline and the material form of images: An analysis of scientific visibility, *Social Studies of Science* 15(1):37–66.

Lynch, M. (1985b). *Art and artifact in laboratory science: A study of shop work and shop talk in a research laboratory*. London: Routledge and Kegan Paul.

Lynch, M. (1985c). La rétine extériorisée: Sélection et mathématisation des documents visuels. In B. Latour and J. de Noblet (Eds.), *Les "vues" de l'esprit*. Special issue of *Culture Technique* 14 (June).

Maynard Smith, J. (1985). The Birth of sociobiology. *New Scientist* (26 September):48–50.

Myers, G. (1986). *Narrative and interpretation in the sociobiology controversy*. Paper presented at the History and Philosophy of Science Seminar, University of Leeds, October.

Myers, G. (Forthcoming). *Writing biology: Texts in the social construction of science.*

Pfeiffer, J. (1975). Review of *Sociobiology. The New York Times Book Review* (27 July):15–16.

Pickering, A. (1988). Editing and epistemology: Three accounts of the discovery of the weak neutral current. In L. Hargens, R.A. Jones, and A. Pickering (Eds.), *Knowledge and society: Studies in the sociology of science, past and present*, Vol. 8.

Rensberger, B. (1975). Sociobiology: Updating Darwin on behavior. *New York Times* (28 May):1.

Rudwick, M. (1976). The emergence of a visual language for geological science, 1760–1840. *History of Science* 14:149–195.

Schiele, B. (1986). Vulgarisation et télévision. *Information sur les Sciences Sociales/Social Science Information* 25:189–206.

Silverstone, R. (1985). Framing science: The making of a BBC documentary. London: BFI.

Tilling, L. (1975). Early experimental graphs. *British Journal for the History of Science* 8:193–213.

Wilson, E. (1975a). *Sociobiology: The new synthesis*. Cambridge, MA: Harvard University Press.

Wilson, E. (1975b). Human decency is animal. *New York Times Magazine* (12 October)38–50.

Wilson, D.O. (1977). Biology and the social sciences. *Daedalus* 106(4):127–140.

Wynne-Edwards, V.C. (1976). Bull's-eye of sociality. *Nature* 259 (22 January):253–254.

Lists, field guides, and the descriptive organization of seeing: Birdwatching as an exemplary observational activity

JOHN LAW
Department of Sociology, Social Anthropology and Social Work, University of Keele, Keele, Staffs ST5 5BG, UK

MICHAEL LYNCH *
Department of Sociology, Boston University, 96–100 Cummington Street, Boston, MA 02215, USA

> Warblers so puzzled the Cherokees that they left many species without names, apparently because they could not tell one from the other – a touch of nature that makes them kin to modern bird watchers who, looking the birds up in their field guides, find them lumped together in the category "confusing" warblers. Joseph Kastner, *A World of Watchers* (1986:7)

1. Aspect-blindness

For most of us, most of the time, the activity of seeing and naming objects in the natural environment is relatively unproblematic. So "natural" does it seem that the contextual skills that we deploy are concealed from our scrutiny. It is only when we are novices – young children, apprentice scientists or radiographers, or aspirant birdwatchers – that the fact of those skills, and, more important, of their social construction, becomes visible to us. Consider these notes which are taken on a birdwatching trip:

* We are grateful to Bob Anderson, Jeff Bowker, Michel Callon and Bruno Latour who read and commented on an earlier draft of this paper.

N is a complete novice, initially with indifferent interest in the activity of birdwatching, but lately taking some interest. She had not "studied" the field manuals, and has not accumulated a fund of previous experiences other than of the "usual" backyard varieties and a few of the more spectacular species seen alongside the highway or on nature walks.

Again and again I point to a specimen, reciting my judgement of its identity: a gadwall, a night heron, black tern, kestrel, etc. I indicate or describe the specimen's locale, she trains her binoculars on it, and I make some remark about its features, habits, abundance, etc. A day, or even an hour later, she'll ask me, "What's that?" and I'll note that it's the same species I had described to her on a prior occasion.

She started keeping a list of her own. She kept asking me whether she should "count" species that I knew she had seen, but that she would be unlikely to recognize again should she see one. She acknowledged that she "didn't know what she was seeing". The question is: what does this "not knowing" consist of?

Assuming that over the course of the occasion we have just successfully collaborated on isolating the same bird (just this duck), the question then becomes, "just what has the name, e.g., 'gadwall' been attached to?" Clearly it's attached to the duck. What else? We didn't misfire and attach the name "gadwall" to a clump of reeds, a pattern of reflection on the water, the insect hovering overhead. It aimed at the duck, but somehow it didn't "stick". The next time N sees what I take to be the identical species she doesn't see a gadwall.

If N is "failing", then numerous possible explanations for her "failure" come to mind: she has failed to recognize that the second duck was also a gadwall because of variations in perspective or individual condition; she recognized that it was like the other one but she forgot what it's name was; although she "saw" the duck she was momentarily distracted from the game of birdwatching and made no effort to identify it by species. None of these explanations are implausible. However, the explanation that we prefer is that, for her, "gadwall" is one name among many for "ducks". What is happening is that, though she can see and name generic ducks, she is unable to see the features of the gadwall as contrasting with other ducks of similar profile: the dull grey appearance and the male's black posterior are not juxtaposed against the green and ruddy heads, colourful wing and flank patches, distinct beak shapes and darting movements and seasonal and regional habitats of comparable species.

Let us say, then, following Wittgenstein, that she is experiencing "aspect blindness".[1] That is, she is "suffering" not from a defect of eyesight or an inability to see or optically resolve birds in the field, but rather from *an inability to collect and re-collect species identifications*. She has not, that is, worked through a *table of possibilities* in the light of a sighting of a duck to find how it differs from others in a taxonomic array.

Where would such a table of possibilities come from? The persistent birdwatcher might create it for herself. This would be unusual and we will consider the possibility no further. She might derive it from another more experienced birdwatcher. She might derive it from the discriminations that are made in a field guide. And/or, she might start creating her own list of the birds she had observed. Typically she would mix the last three. However, for the purposes of simplicity we will concentrate primarily on lists and field-guides and describe a particular "literary language game"[2] *in which a "novice" walks or drives through a* particular habitat and consults a field manual as an aid to formulating a list of birds seen on the expedition.

The notion of a literary language game brings into relief the way in which naturalistic observation and representation require an apprenticeship in a social organization of "reading" and "writing". When encountered through such an apprenticeship, "natural order" is discovered and organized through the basic texts in the language game. "Natural kinds" are not simply representations of what the eye (or the mind's eye) sees. In place of this perceptual model for observations we are substituting a model of *reading and writing*. We are suggesting that birdwatchers do not simply see birds. Rather: they (1) engage in a reflexive elaboration in which a text provides an iterable organization, a bulky object and a moment in a hermeneutic reading of the world; and (2) organize their gaze sequentially, in terms of the canonical order of a list. If these suggestions can be sustained for the accessible activity of birdwatching, they may also be located in esoteric practices in the natural sciences.

2. The descriptive organization of seeing: Lists

> He binds sight to crafted description and, further, places this activity in the context of the greater Baconian project. Svetlana Alpers, *The Art of Describing: Dutch Art in the Seventeenth Century* (1984:73)

Lists are central to birdwatching, and practitioners commonly keep several: "life lists" which record a cumulation of species identified by a particular person; lists which record species identified in a particular time interval; lists of species identified on a particular trip by an individual or group; and "Christmas counts" which record the species identified within a particular regional jurisdiction during an organizationally specified time period. Superficially, such lists are records of the species observed by members: they are representations of observations. However, they are also much more than that. How any observation is organized in the course of a field trip depends upon the lists being compiled in-and-through the observation. "Perception" is list-driven in the sense that the current state of the list provides motives for: searching the environment; regarding, disregarding and selecting among potential experiences; remarking upon or saying nothing about an observed event; and treating an announced sighting as a notable, doubtful or unremarkable claim. There is thus a reflexive relation between the literary phenomenon of the list and the embodied and interactional performance of observation and representation. The concrete configuration of observation is not reducible to generic structures of individual perception and cognition because it depends upon the textual formatting of a list and the source of social interaction through which that list is composed.

Although a solitary birdwatcher may keep a list for herself alone, list making and list reciting are organizationally accountable. Birdwatchers can be competitive about the length of their "life lists", and listings of uncommon or rare species are subject to rivalry and controversy (Kastner, 1986:211ff.). Organizations such as local chapters of the Audubon society, which collate sightings of rare and vagrant species in a region, often specify constraints on what is to count as a competent identification. As the following practical advice suggests, to identify a bird properly (i.e., to *list* it in a socially acceptable way) is to build a canonically ordered description:

> Notes should be made and kept in logical and systematic sequence, if possible, for ease of later retrieval. Try to build a description each time in the same order. Do this by looking for different parts of the bird in the same order. This is, of course, not always possible and often you must scramble and take what you can get when you can get it but *trying* to follow the same sequence is a start at learning a good habit. What is more important is writing up the details. (Bernstein, 1984:1)

In this ideal world the bird watcher is supposed to start by noting her overall impression of the bird – its "feel": "Study the bird in life to its gestalt, that is, the bird in its entirety." (Bernstein, 1984:1) Then she is instructed to move through a list of parts in a specific order – starting with the dorsal and moving through the ventral and the "soft parts" (the eye, the mouth, the feet) to a description of its song, if any. In this particular article a schematic picture of a generic gull (the author notes this to be "admittedly oversimplified") accompanies the text and points to the different types of feathers which cover the wing. This "logical and systematic" check list and the description that it shapes is contrasted with that which may be derived from the field manual:

> Compare your sighting with books only after the notes are made. Having the book at hand during the note-taking will only interfere with the process. Many possibly good and valid records have been tarnished because the observer consulted a book before finishing the notes. As a result, the description often is that of the picture in the book, *not* of the actual live bird seen. (Bernstein, 1984:2)

Such an attempt to impose a standardized form by those collating bird lists is not surprising. Thus it is only when descriptions are normalized in this way that it becomes easy to compare descriptions in the search for authoritative similarities and differences between species.[3] While Garrett argues that an "accurate and detailed description" is always necessary, he also notes that the:

> documentation of a rarity must convince reviewing bodies or editors that similar species were considered and reasonably eliminated. Having in mind the species one needs to eliminate, one becomes selective about the aspects of the bird under scrutiny which are emphasised in the description. (Garrett, 1986b:3)

He goes on to note that recording committees have to "juggle multiple descriptions" with different or conflicting interpretations.

Lists, we suggest, answer to the problems of imposing durability upon bird sightings. Inserting such sightings from a local context into a network which monitors, selects and compiles lists on a collaborative basis depends upon their translation into a canonical and normalized form in which one set of textually expressed similarities and differences is emphasized at the expense of others.

3. The descriptive organization of seeing: Field guides

In North America or Europe the birdwatcher may choose from a range of field guides covering national, geographical and local regions. Popular field guides have been used for at least two centuries, but the "classic" guide currently in use in North America is Peterson's *A Field Guide to the Birds of the Eastern United States* (1934; 1939; 1947; 1980). Peterson's Western Edition (1941; 1961) is similar in format to the original Eastern Edition, and is the main source for our analysis of Peterson's illustrations and descriptions. Original and revised editions of Peterson's work remain in widespread use, although several other popular field guides have appeared in recent years. Two of these, which we shall compare to Peterson's, are the Audubon Society guide (Udvardy,

1977) which uses photographs rather than drawings of birds, and the National Geographic Society (1983) guide which, according to Kastner (1986:208), is aimed at more "sophisticated" bird-watchers.

Though these guides display important differences, they also have a number of points in common:

1. *Naturalistic accountability*: all three field guides are realist in some sense, though each exhibits a different accent. Each manual, and its novice user, operates on a set of commitments: that bird species exist in nature; that they can be identified and indexed on the basis of sensory (mainly visual, but also audible) evidences; that separate species can be identified and named; and that species can be represented in paradigmatic illustrations and described in texts. The entire literary language game relies upon and testifies to these naturalistic assumptions.
2. *Authority*: when playing the literary language game, the novice relies upon the text as a "state of the art" compendium – as an authority and a "disciplinary matrix" in Kuhn's (1977a:307ff.) sense which lists each of a region's available species. Such authority may not be entirely justified. Classification systems have changed historically, and continue to change, as witnessed by recent changes in the "lumping" together of the Bullock's and Baltimore orioles into the category of Northern oriole, the imperialistic absorption of the Florida gallinule by the common moorhen, and the annexation of four species of juncoes into one species (Kastner, 1986:213–214). Many manuals are notably incomplete in their listings of species, and no field guide lists every species or variant that might *possibly* be seen. Nevertheless, however unjustified this may be, the text remains authoritative in the hands of the novice unless strong external grounds are found for denigrating its completeness or adequacy. Efforts are made to "locate" a bird sighted in the field in the text's listings of pictures and descriptions.[4]
3. *A picture theory of representation*: each of the three field guides is lavishly illustrated. A non-illustrated field guide would be almost impossible to use in the field. Each guide thus employs a tacit "picture theory" of representation: an idealiza-

tion of the potential correspondence that can be achieved between a representation in the text and the "bird in the field". The illustrative conventions, however much they may differ in the schematic or photographic form, are taken (within the language game) as realistic in *some* sense. The question is: what is that sense? Their realism is conveyed by "illusionist" (Gombrich, 1960) use of the pictorial surface to provide a realistic sense of what can be seen in the field. While they may also count as aesthetically appreciated commodities in themselves, within the novice's literary language game the pictures act as mundane referential devices.

4. *A strategic use of texts*: each manual makes strategic use of captions and descriptions. A wordless picture book for birds – where pictures were presented without identifying names, captions, pointers, range maps, phonetic spellings of birds-calls etc. – would be useless in practice. The interplay of pictures and other expressions in each manual furnishes it's readable relation to birds in the field. In addition, various forms of index are used in the field guides to facilitate the sometimes hopeless task of finding the appropriate pictures under the severe time-constraints of identifying a bird before it flies away.

3.1 Schematic, photographic and dioramic birds

Despite these important similarities, the three manuals nevertheless take very different representational paths. Peterson's early and revised editions are the most schematic, the Audubon guide opts for photographic realism, and the National Geographic guide uses somewhat more naturalistic paintings than does Peterson.[5]

3.1.1 Peterson's schematic representations

In Peterson's various editions colour and black and white plates (see Figures 1 and 2) juxtapose representative drawings of several different species on a single page. Each species depiction is arrayed as a decontextualized specimen (usually juxtaposing male, female and/or immature instances in a cluster), drawn in paradigmatic fashion and with a minimum or absent reference to naturalistic surroundings. Each drawing is labelled, and a small line indicates the feature or features of the bird in question that are held to be particularly relevant in assisting identification. On any single page each representative bird is drawn in parallel orientation to others on the same page so that the various specimens make up a loosely tabular arrangement. For some specimens – for instance the hawks – separate arrays are used to depict flying profiles as seen from underneath. In the earlier Peterson editions (Figure 1) the perching plate is faced by a sparse page of text which lists the pictured birds within their genera, and comments on the features indicates by the lines on the drawings. A fuller description is provided elsewhere in the texts. In the most recent edition (Peterson, 1980) the description is placed opposite the plate, and a distribution map is separately printed toward the end of the guide.

Peterson is very clear that the drawings in both guides are stylized. It is worth quoting him on this subject at some length:

> The plates and cuts throughout the text are intended as diagrams, arranged so that quick, easy comparison can be made of the species that most resemble one another. As they are not intended to be pictures and portraits, modelling of form and feathering is often subordinated to simple contour and pattern. Some birds are better adapted than others to this simplified handling, hence the variation in treatment. Even color is sometimes unnecessary, if not, indeed, confusing. (Peterson, 1947: xviii)

In the most recent edition he writes that:

> Because of the increasing sophistication of birders I have leaned more toward detailed portraiture in the new illustrations while trying not to lose the patternistic effect developed in the previous editions. A drawing can do much more than a photograph to emphasize the field marks. A photograph is a record of a fleeting instant; a drawing is a composite of the artist's experience. The artist can edit out, show field marks to best advantage, and delete unnecessary clutter. He can choose posi-

Plate 16

HAWKS

ACCIPITERS (True Hawks)
Small head, short wings, long tail.
Adults, barred breasts; immatures streaked.

COOPER'S HAWK — p. 50
Crow-sized; rounded tail.

GOSHAWK — p. 49
Adult: Pearly breast, light gray back.
Immature: Large; pronounced eye-stripe.

SHARP-SHINNED HAWK — p. 49
Small; notched or square tail.

FALCONS
Large head, broad shoulders.
Long pointed wings, long tail.

SPARROW HAWK or AMERICAN KESTREL — p. 61
Both sexes: Rufous back, rufous tail.

PIGEON HAWK or MERLIN — p. 60
Male: Small; slaty back, banded gray tail.
Female: Dusky back, banded tail.

PEREGRINE FALCON — p. 60
Adult: Slaty back, light breast, black "mustaches."
Immature: Brown, streaked; typical "mustaches."

HARRIERS
Small head, long body.
Longish wings, long tail.

MARSH HAWK (HARRIER) — p. 57
Male: Pale gray back, white rump.
Female: Brown, with white rump.

Figure 1. Peterson's schematic hawks

Illustrations and captions from the book, *A Field Guide to Western Birds*, written and illustrated by Roger Tory Peterson, published by Houghton Mifflin Company, Boston. Copyright 1941, 1961 Roger Tory Peterson. Reprinted with permission.

tion and stress basic color and pattern unmodified by transitory light and shade. ... The artist has more options and far more control .... Whereas a photograph can have a living immediacy a good drawing is really more instructive. (Peterson, 1980: 9–10)

Plate 17

ACCIPITERS, FALCONS, KITES, HARRIER

ACCIPITERS have short, rounded wings, long tails. They fly with several rapid beats and a short glide.

COOPER'S HAWK p. 50
Near size of Crow; rounded tail.
GOSHAWK p. 49
Adult: Very large; pale pearly-gray breast.
Immature (not shown): See text.
SHARP-SHINNED HAWK p. 49
Small; tail square or notched.

FALCONS have long, pointed wings, long tails. Their wing strokes are strong, rapid but shallow.

PEREGRINE FALCON p. 60
Falcon shape; near size of Crow; bold face pattern.
PRAIRIE FALCON p. 59
Dark axillars (in "wingpits").
APLOMADO FALCON p. 60
Black belly, light chest.
PIGEON HAWK or MERLIN p. 60
Banded gray tail.
SPARROW HAWK or AMERICAN KESTREL p. 61
Banded rufous tail.
GYRFALCON p. 59
Larger than Peregrine; grayer, without contrasting pattern. Arctic. A white phase also occurs in our area.

KITES are falcon-shaped but are buoyant gliders, not power-fliers.

WHITE-TAILED KITE p. 48
Falcon-shaped, with white tail.
MISSISSIPPI KITE p. 49
Falcon-shaped, with black tail.
Immature: See text.

HARRIERS are slim, with somewhat rounded wings, long tails, and long bodies. They fly low with a vulture-like dihedral and languid flight.

MARSH HAWK (HARRIER) p. 57
Male: Whitish, with black wing-tips.
Female: Harrier shape; brown, streaked.

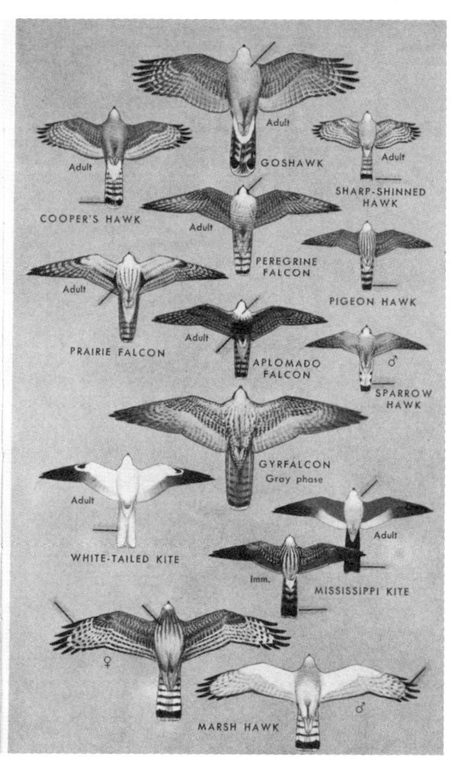

Figure 2. Peterson's schematic hawks, flying

Illustrations and captions from the book, *A Field Guide to Western Birds*, written and illustrated by Roger Tory Peterson, published by Houghton Mifflin Company, Boston. Copyright 1941, 1961 Roger Tory Peterson. Reprinted with permission.

Peterson's view that we should treat the illustrations as instructive diagrams to bring out authoritatively recognizable similarities and differences, rather than as full representations – can be exemplified in his treatment of two similar species – the Sharp-Shinned Hawk and the Cooper's Hawk (see Figures 1 and 2). Figure 1 shows two and three drawings of each species in equivalent postures and may be thought of as a table designed to contrast what might otherwise be thought of as indistinguishable species. Thus the profiles are placed along a vertical dimension, a column. Or perhaps, since there is no heading separate from the arrangement of the profiles, it would be better to say that the fact that there is a vertical dimension is made apparent in the way in which the illustrations are arrayed. This tabularization is assisted by the almost complete deletion of what Peterson describes as "clutter".

Let us consider some of these deletions. The first of these concerns distance. Thus, although none of the illustrations use the conventions of linear perspective (chiaroscuro, vanishing points, etc.), it is clear that Peterson makes use of depth in the quasi-tabular arrangement. We take it that the slight relative differences in size between the picture of the two types of hawk are not irrelevant but rather indicate the relative sizes of the birds. To see this, the reader must assume that each set of specimens is calibrated for "distance" from the reader and that the flattened depiction is not simply a non-perspectival rendering, but acts, rather, to "control the variable" of distance. This reading is partially warranted by reference to the text which (in the case of the more recent guide) indicates that the male Cooper's Hawk is "obviously larger" though it also indicates that there may be size overlap between a female Cooper's and a male Sharp Shinned Hawk.

Distance is not, however, the only visibly constructed *ceteris paribus* relation. Though there are substantial differences in the colouring of the two species *between* the two guides (the earlier is "richer" in tone), *within* each guide the colouring conventions adopted for each bird are identical. This similarity may be compared with the photographs in the Audubon Society guide (Udvardy, 1977) where the wings of the Cooper's Hawk appear to be darker and browner than the greyer Sharp Shinned specimen (Figure 3). Again, in the Audubon guide, in so far as it is possible to tell (the bird is facing away from the camera), the breast of the

Figure 3. Photographs of hawks

Illustrations from *The Audubon Society Field Guide to North American Birds, Western Region*, edited by M.D.F. Udvardy, published by Chanticleer Press, Inc., New York. Plate 324 by Harry Darrow, Plate 325 by Karl Maslowski, Plates 326 & 327 by Ron Austing. Reprinted with permission.

Sharp Shinned also seems to be lighter than that of the Cooper's. Colour, then, is a second area in which Peterson controls for what he takes to be irrelevant clutter. If the diagrams may be seen as instructions, then they are telling the reader who wishes to distinguish between the two species not to attend to the colouring of the birds.

A third visibly constructed *ceteris paribus* relation has to do with posture and orientation. As we have indicated, Peterson offers the reader identically posed depictions. These may be compared with the very different photographs offered of the two species in the Audubon guide – a Sharp-Shinned on the wing taken from underneath, an immature Sharp Shinned perched with its back to the camera, and a Cooper's, also perched, but half facing the camera. Though few birdwatchers would read the Audubon guide in this way, these differences could be used to legitimate the assumption, for instance, that mature Sharp Shinned Hawks are more likely to be in the air than mature Cooper's. Less fancifully, the Peterson drawings may also be contrasted with his pictures of Chickadees, Nuthatches and Brown Creepers.[6] There, where posture is a vital clue to the detection of difference, it is indeed represented in his drawings.

Peterson effects a final visual construction of *ceteris paribus* by adopting the already mentioned convention of pointing, by use of lines or arrows, to the visual differences between species that are authoritatively held to be crucial in difficult cases of identification. Thus, in the case of the two Hawks, it is the shape of the tail which is held to be particularly important for visual diagnosis. The instruction which may be read from the use of these lines once again, then, amounts to a *ceteris paribus* clause: if the birdwatcher has succeeded in narrowing down the identification of the bird in question to the point where it is either a Cooper's or a Sharp-Shinned, then all other visual differences may be ignored in favour of the shape of tail.

Overall, Peterson's drawings and their arrangement facilitate a number of inferences: first, the proximate and tabular arrangements of the two species eases their comparison. We are being asked to see these two species as close to one another, and their comparability is highlighted by their proximate and parallel illustration; second, visual control is imposed on all features save the "relevant" variables of size and tail shape. If these, rather than, say, colour and posture, are to stand as criteria for difference, then they do so in part because all other visible differences have been ruled out in the formal structure of the illustration. The comparative ambiguity of the Audubon book (discussed below) allows us to reflect on how Peterson provides clear criteria by artfully rendering most possible differences irrelevant.[7]

The texts in Peterson reiterate many of the relations that are *shown* in the tabular depictions. The verbal accounts of the two Hawks follow one another. The lists of traits (size, colouring, shape of wings, tail, etc.) for each are similar, except for italicized differences: "generally the Cooper's has a *rounded* tail (Sharp-shin *square-tipped* tail, slightly notched when folded)" (Peterson, 1961:50). These differences elaborate on the instructions which can be derived from the drawings, though they also, in the case of the two birds under discussion, tend to add complication:

> It can be very tricky separating small male Cooper's Hawks from large female Sharp-shins. They are not much different in size and the Sharp-shin's square-tipped tail can even look slightly rounded when spread fanwise. The tail shape works best when the tail is folded.[8] (Peterson, 1947:42)

The recent edition frankly admits that "many cannot be safely identified in the field" (Peterson, 1980:152). Other differences which cannot be *shown* are described – for instance the call of the two species, and their range.

Note that the described similarities and differences do not simply criss-cross the various textual sites and formats but also make use of extrinsic bases of identity and comparison. The Cooper's is described as "not quite as large as" a Crow (Peterson, 1961:50), and the Sharp-Shinned is differentiated from "other small Hawks" by its short rounded (rather than long pointed) wings. Thus, while the field guide is written to instruct the novice, a minimum degree of assimilation to the work of birdwatching is assumed. A number of "common or garden" species are used as reckoning points for the identification of others, and the network of similarities and differences depends upon and elaborates an increasingly esoteric knowledge of species as competence develops. Thus, while various orders of common sense are assumed on the part of readers (familiarity with language, maps, sounds and sights, habitat types), the use of indicator species posits the existence of a "core" of commonplace skills for recognizing and categorizing bird species. Robins, sparrows, and crows are assumed as "known" species by "anyone" who might use these guides.[9]

3.1.2 The Audubon Society guide's photographic realism

As we indicated earlier, the Audubon guide opts for photographic realism (see Figure 3) and uses actual photographs of representative specimens in its colour plates. The author defends this decision by arguing that:

> photographs add a new dimension in realism and natural beauty. Fine modern photographs are closer to the way the human eye usually sees a bird and, moreover, they are a pleasure to look at. (Udvardy, 1977:10)

Two photographs are placed on any single page. Typically, a single bird is shown in the frame of any photograph, and one specimen of an adult and one of an immature bird is used for each species. For some, for instance the Sharp-Shinned Hawk, a photograph of a flying as well as a perching specimen is shown. Each photograph is labelled, the size of the bird is indicated, and a page reference to a written description is given. Plates are indexed by generic profiles, sometimes colour-coded for the dominant hue of the bird (red, yellow, brown, etc.).

The photographs, though they juxtapose closely similar species, and appear, when taken in conjunction with the text, to be interpretable as instructions about what should be attended to when identification is attempted, are, for the reasons mentioned above, otherwise unlike the drawings in the Peterson guide. The version of photographic realism used in the Audubon guide less clearly highlights the similarities and differences between species. First, it is difficult if not impossible to read the photographs as tables. The birds appear in a variety of poses and orientations. Second, detail treated by Peterson as clutter is here not deleted. The birds appear against a background of branches, leaves or sky. Taken together, these factors lead to a third consequence: it is difficult if not impossible (a) to notice and (b) to interpret the significance of variations in the size of images: it is necessary to refer to the caption to discover whether there is a difference in size between two birds. In addition, the provision of a background offers rich detail whose salience is, however, not easily interpretable. Thus, the possibility that the Sharp-Shinned Hawk frequents deciduous trees while the Cooper's prefers conifers, while a warrantable reading of the background of two of the photographs, is undermined by the discovery that the text about the Cooper's is located in the section on deciduous birds, while the Sharp-Shinned is to be found in the section on conifers! Fourth, the colouring of the birds, and complexities of feathering are not stylized, and the photographs thus do not, as Peterson prefers, "emphasize the field marks". Indeed, as we have already indicated, differences in colouring between the perching Sharp-shinned and Cooper's Hawks which are bleached out of Peterson's account, are quite noticeable in the Audubon guide. Since the text of the latter writes that the Cooper's male is "slate blue above, barred rusty below" (Udvardy, 1977:638) while the Sharp-Shinned is "Slate blue above, white below with rich rusty cross-barring" (Udvardy, 1977:697), it is, perhaps, pointing to differences in colouring as an important distinguishing feature. Whether or not this is the case is unclear, given the absence of stylization in the photographs. The multi-interpretability of the illustrations is aided by the fact that there are not, as in Peterson, lines or arrows on the photographs to indicate salient differences which particularly deserve attention.

3.1.3 The National Geographic guide's naturalistic diorama

The National Geographic guide which is the most recently published of the three, and is generally the preferred guide for the birdwatchers observed in this study, combines characteristics of the other two. The illustrations are hand-drawn by various artists (Figures 4 and 5), but are more "naturalistic" in the depiction of positions and surroundings than those in Peterson. On any given colour plate, typically two or three "related" species are depicted. Each species is represented by several individuals (adult male, female, one or more "immatures", and sometimes flying as well as perching specimens). The various individuals of a species are posed in "natural situations", and the background features are "filled in" with somewhat greater perspectival detail than in Peterson. On each page (such as Figure 4), two, three or more such "diorama"[10] are juxtaposed discontinuously with each other and various minaturized profiles of flying or strategically positioned specimens.

The "diorama" suggest a natural situation by detailing the birds perched on branches in an appropriate habitat. The hawks in Figure 4, unlike Peterson's, are shown in the midst of characteristic activity. One member of each pair of hawks on the page is shown with prey. Like a museum diorama, however, the arrangements are discretely framed synthetic schemes. They differ from the Audubon Society's guide by showing, within each of the three frames shown on the page, perching adults of both sexes and a flying immature. The sharp-shinned hawk diorama includes the added bonus of a hawk, shown in the distance, pursuing an evening

Accipiters

Low-flying woodland hawks with short, rounded wings and long tails for speed and agility. Females are larger than males. Three species occur in North America.

Sharp-shinned Hawk *Accipiter striatus*

L 10-14" (25-36 cm) W 20-28" (51-71 cm) Distinguished from Cooper's Hawk by shorter, squared tail, often appearing notched when folded, and smaller head and neck. Adult lacks strong contrast between crown and back. Immature is whitish below with bold, blurry, reddish streaking on breast and belly; narrow white tip on tail; head less tawny than in Cooper's or Northern Goshawk. In flight (see also page 207), note smaller head and proportionately short tail and long wings. Fairly common over much of its range; found in mixed woodlands. Migrates mostly along ridges and coastlines, often in larger flocks than Cooper's or Goshawk.

Cooper's Hawk *Accipiter cooperii*

L 14-20" (36-51 cm) W 29-37" (74-94 cm) Distinguished from Sharp-shinned Hawk by longer, rounded tail, larger head, and, in adult, stronger contrast between back and crown. Immature has whitish or buffy underparts with fine streaks on breast; streaking is reduced or absent on belly; white undertail coverts; white tip on tail broader than on Sharpshin or Northern Goshawk. Tail proportionately longer, wings shorter than Goshawk. In flight (see also page 207), note large head, and proportionately long tail and short wings. Uncommon and may be declining. Inhabits broken woodlands or streamside groves, especially deciduous. Preys largely on songbirds, some small mammals. Sometimes perches on telephone poles, unlike Sharpshins. Migrates mostly along ridges and coastlines, usually singly or in groups of two or three.

Northern Goshawk *Accipiter gentilis*

L 21-26" (53-66 cm) W 40-46" (102-117 cm) Conspicuous eyebrow, flaring behind eye, separates dark crown from blue-gray back. Underparts are white with dense gray barring; appear gray at a distance. Adult has conspicuous fluffy white undertail coverts, more extensive than in Sharpshin and Cooper's. Immature is brown above, buffy below, with dense, blurry streaking, heaviest on flanks. In all ages, tail has wavy bands and a thin white tip; in immature, dark bands are bordered with white. In flight (see also page 207), note proportionately short tail, long wings. Sometimes confused with Gyrfalcon (page 204). Fierce and bold even for an accipiter, the Goshawk inhabits deep, conifer-dominated mixed woodlands; preys chiefly on ground-dwelling birds and ducks, also on other birds, and mammals as large as hares. Uncommon; winters irregularly south of mapped range in east. Migrates mostly along ridges and coastlines; southward irruptions occur in some winters.

190

Figure 4. Hawks in diorama

Illustrations and captions from *Field Guide to the Birds of North America*, published by the National Geographic Society, Washington D.C. Drawings by Donald L. Malick. Reprinted with permission.

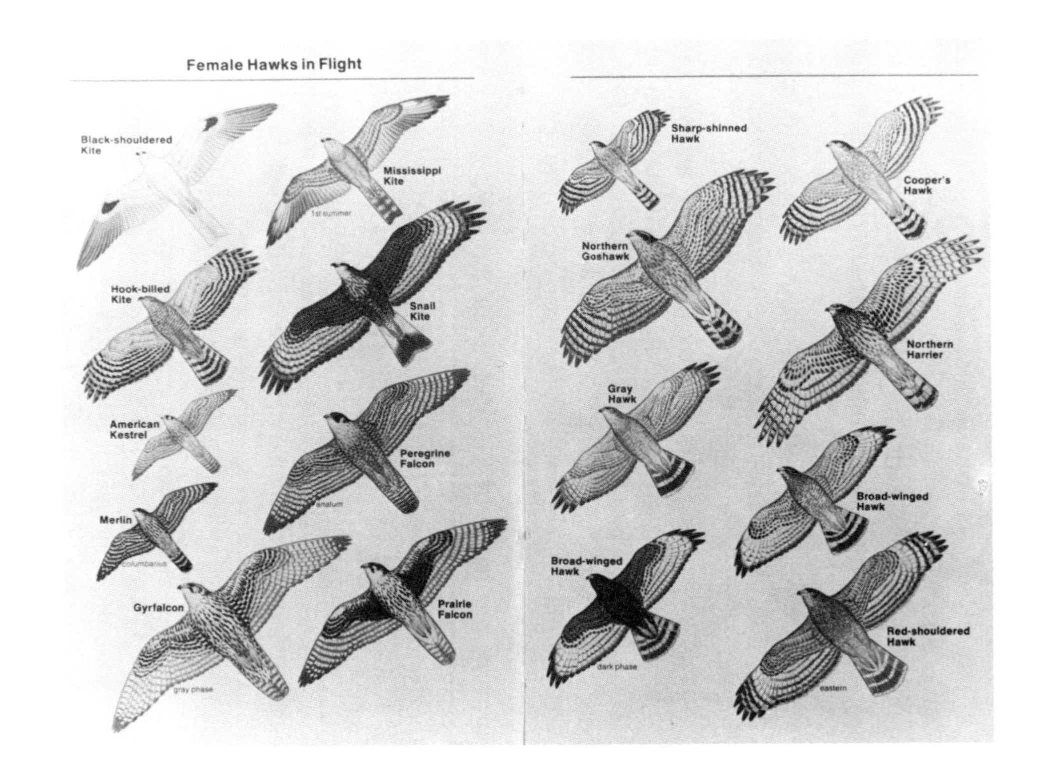

Figure 5. Flying hawks, National Geographic guide

Illustrations from *Field Guide to the Birds of North America*, published by the National Geographic Society, Washington D.C. Drawings by Kent Pendleton. Reprinted with permission.

grosbeak. Although individuals of each species are not represented in exactly the same pose as in Peterson, the specimens are arranged in such a way as to minimize overlap, while packing the various individuals into a compact space. Each adult specimen displays eye, beak, a portion of breast, tail, and feet. And, while each of the three frames is isolated from the others on the page, the specimens are shown in proportionate size. The "controls" in this case are far less obtrusive than in Peterson, but no less significant. While evoking the detail of a photograph, the frame includes an *organization* of such detail that would be extraordinary if found in any single photograph.

Like Peterson's guide, the National Geographic uses a tabular array of flying hawks where the specimens are calibrated for apparent "distance" and are oriented in a parallel direction, with "cognate" specimens juxtaposed (Figure 5). There are several differences, however. The interior of each silhouette shows more profuse detail than in Peterson, with the outlines of the flight feathers and their barred patterns more distincly painted. The National Geographic's hawks look less like cardboard cutouts than Peterson's, since their partial side view, showing of eye and facial features, gives more of an illusion of dimensionality. Although this does not show up in our black and white reproductions of the figures, Peterson's array of flying hawks (Figure 2) is done in black and white, while the National Geographic's is in color. Whether this makes them more "realistic" is questionable, since if "realism" refers to what a viewer sees when inspecting a flying hawk with binoculars, the highly detailed National Geographic hawks are not very realistic. From below, in a bright sky, a birdwatched typically sees a back-lighted, obscurely marked, silhouette.

The National Geographic guide makes no use of Peterson's famous lines to point out field marks, although in its descriptive caption for the Sharp-Shinned hawk (Figure 5) it does mention, among a number of other distinctive features, a "shortened, squared tail, often appearing notched when folded". Generally, the National Geographic guide places less emphasis on single field marks, opting for a list of visible and behavioral features (e.g., "sometimes perches on telephone poles, unlike Sharpshins").

A distinct advantage of the National Geographic guide is that it presents picture, name, description and range map all on a single, two-page spread. This differs from Peterson's early editions where extended descriptions are placed on a separate page from the illustrations, and from Peterson's most recent edition (1980), where longer descriptions are placed alongside the illustrations while range maps are indexed to a separate page. It also differs from the Audubon Society Guide, which places its colour plates in a separate section from pictures and range maps. The National Geographic guide enables the reader to more quickly assemble the various resources the text supplies for identifying a given species.

4. The circumstantial deconstruction of textual order

However well organized the field guide and however admirable the programmatic aims of its author, our experience with all three of the manuals discussed above convinces us that users of the guides will encounter innumerable frustrations, uncertainties and quandries. Such "troubles" are typically experienced by committed birders as temporary problems arising within a personal and situational relationship to "reality" – problems with perspective, acuity and luck. While we admit that a "positive" attitude may encourage noble efforts to overcome difficulties, we see such troubles as being deeply embedded in the organization of the novice's literary language game. As such, they have methodological significance as "pauses" in the midst of fluent practice which reflexively call forth taken-for-granted organizations of activity. We therefore embrace such negativity in order to gain leverage on the texts that we are analyzing, and to show how their "reading" is caught up in contingencies which cannot be found in even the most detailed inspection of their pages.[11]

We will start by considering troubles that are associated with the particular organization of each of the field guides discussed above. Our concern is not to evaluate the quality of the guides themselves but rather to show how the representational conventions which give each text its distinctive advantage as a field guide can also be sources of trouble. We will then move to those troubles that may arise in the use of any manual.

4.1 Troubles with particular field guide conventions

4.1.1 Troubles with schematic representations: "Missing the mark"

This arises most clearly in the case of Peterson's treatment, with its emphasis on one or a few particular "field marks" for each species, and its formalistic "bleached out" illustrations. The other two manuals are not immune to this trouble since the Audubon guide's photographs are selected to show birds in paradigmatic poses, and the descriptions emphasize key distinguishing features. Similarly, the naturalistic paintings in the National Geographic guide highlight certain features and the texts list particular distinguishing marks. The trouble is the following. Often the bird flies away even before a "good look" can be obtained. There are innumerable ways in which this occurs: the bird flies before it can be seen within the narrow field of vision of the binoculars; it fails to expose the parts of itself upon which the "field marks" are detectable; it moves rapidly; it remains too far away.

There is a related problem with all but the most recent Peterson edition. This is because more detailed descriptions are placed elsewhere in the book than the illustrations. As Figures 1 and 2 show, there pictures are paired with names and very brief accounts of key field marks on the facing pages. Often, when using Peterson in practice, we have found the illustrations and brief captions to be insufficient, and have turned to the more extensive descriptions. This takes extra time, assuring in many cases that the yet-to-be identified bird flies away or hides itself. In addition, the Western Edition does not include range maps, and this has proved frustrating at times since often the easiest way to distinguish between similar appearing species (such as the Eastern and Western meadowlark) is to consult the range map. This practice, which is sometimes frowned upon, is rendered most easy by the National Geographic guide.

4.1.2 Troubles with photographic realism

As we have already noted, despite their "realism" photographs can be a source of innumerable "illusionary" uses. Perhaps unsurprisingly, in practice we have found the Audubon guide to be the most frustrating of the three in use. This has little to do with the quality of the photographs but rather with the comparative lack of freedom for the photographer (or editor) to represent paradigmatic instances. Although the photographer no doubt selects a "good specimen" (such as an adult in healthy condition and in breeding plumage) and a shot which "captures" it in a "classic" pose, the artist has much greater licence to show a bird in just the pose necessary to reveal its distinctive colouration (using near cubist licence to free the specimen from particularistic perspective), and to highlight the field marks. In addition, the photographer has much less leeway to "control for" illumination and distance than Peterson, and specimens photographed in the field are rarely found grouped in tight tabular arrangements of individuals of distinct species.

As a result the user of the Audubon guide is faced with the problem of extracting the "essential" details that can aid identification while disregarding other details as "gratuitous". In addition, some distinguishing features may not be shown. The plates in the guide show ducks in water without visible feet, or hawks without visible breast feathers, and because the photographs are large, the number of individuals representative of age classes and colour phases for each species is limited. The problem is compounded by the fact that the pictures are paired only with species names, and an index for the page elsewhere in the book where descriptions and range map are found.

By now it is perhaps clear, at least in part, why the Audubon guide is difficult to use, at least in terms of the conventions that have developed in North American birdwatching where almost every birdwatcher has been weaned on Peterson and his system of stylization and little lines. An authoritative reading of the salient differences between illustrations is difficult if not impossible without complete prior competence – it is, accordingly, difficult to treat the illustrations as unambiguous instructions,[12] though to say this is to say nothing about their aesthetic or technical qualities. (Note that a more recent edition of the Audobon Guide, published since this writing, uses photographs in, perhaps, a more convenient way.)

Finally, despite its initial plausibility, the reader should not necessarily, in any case, accept Udvardy's claim that "Fine modern photographs are closer to the way the human eye usually sees a bird". This suggestion is at best questionable. First, photography rests upon a set of technical conventions that may or may not – this depends on circumstances – parallel those that help to constitute the literary language game of the novice. Thus, since, as we have just noted, Peterson's earlier guide helped to train the perception of generations of North American birdwatchers, one should at least entertain the possibility that perception seeks out and notices details – for instance, the round shape of the Cooper's tail by comparison with that of the Sharp-Shinned – which are discernable only with difficulty in photographs.[13] Second, photography, with its technical apparatus of films, film speeds, and levels of light, which is both a constraint and a resource, may see things differently – not at all, or "better" – than the birdwatcher. Hosking's justly famous sequence of photographs of a Barn Owl represents just such a case. Only photography with flash could have captured the owl in the act of bringing a vole to its nest in the dark (Hosking, 1979). And even in the less esoteric case of the photograph of the immature Sharp-Shinned Hawk, there is some evidence that the photographers' art is at work because the lighting in that photograph appears to be artificial, at least in part. Third, the fact that the photographer freezes a moment in the life of a bird creates considerable potential distance between her vision and that of the birdwatcher. First, the moment which has been captured may or may not be consistent with the behaviour of that bird as this is accepted by convention. Second, many of the most characteristic visual features of species are movements, or sequences of movements, which are not captured either by photography or in drawings. And third, it is, in any case, a rare bird that sits still and allows itself to be studied in the way it is possible to study a photograph. Thus, though it would be a project in its own right, there are excellent grounds for questioning both the "realism", and the correspondence theory of perception, which underlie the assumption that photographs are "more true to life" than drawings.

4.1.3 Troubles with naturalistic diorama

Though the National Geographic guide has some advantages over the other two, it also generates troubles for the user. Consider the resemblance between the National Geographic guide's diorama and John J. Audubon's famous paintings in *Birds of America*. Both provide lavishly detailed colour portraits of groups of birds in idealized "situations". It is not incidental that Audubon, like most nineteenth-century naturalists, used a shotgun to collect his specimens so that he could then gain access to their detailed features and pose them more freely than the living specimens might pose themselves. The naturalistic diorama invites the reader to appreciate a "natural" scene, with little acknowledgement of the artifies of pose, position and highlighting. Consequently, readers may be led more easily to expect to see birds look "just like" those portrayed in the guide, and then to suffer a form of "disillusionment" when finding that birds are not quite so colourful, unruffled, clearly marked and boldly positioned.

4.2 Ubiquitous troubles

4.2.1 Troubles with the actual bulk of the manual

Often disregarded in academic treatments of texts is the fact that texts are weighty artifacts, that they require "manual handling", and that pages must be turned in a time-consuming process to find "places" in the text. Birdwatchers become acutely aware of such properties. Typically a text is carried along with binoculars, and must either be held in the hand or stuffed in a pocket or a backpack where it can be instantly accessible. The larger (and thus more "complete") the manual, the more trouble it thus presents to its bearer. In addition, sighting a bird with binoculars often requires both hands to steady the instrument. This makes it difficult or impossible to keep the field guide in the hand. Birdwatchers often wish they were endowed with three arms – two for the binoculars and one for the field guide – and those who wear glasses feel the need for four.

4.2.2 Troubles with place-finding in the text

In addition to the above-mentioned problems, the novice experiences complex difficulties once an interesting but unidentified bird is sighted. Consider a typical instance. A "hawk-like bird" spotted overhead with the naked eye is scrutinized with some difficulty through the binoculars under adverse lighting conditions in order to detect its markings.[14] The novice then turns to the field guide to find a possible identity for the bird, or to confirm what she believes it to be. The necessity of holding the field guide, opening it, flipping through its pages, and reading it under such circumstances, provides a convincing lesson on how the manual is anything but a transparent medium of representation. Binoculars must be put aside, a pause in the action must be negotiated at a delicate time, occupying both hands and eyes. The specimen may fly away or land behind opaque foliage while the reader's gaze is directed to the text. In short, it may refuse to wait until the reader finds the section for "hawk-like birds", checks the pictures and descriptions for several species of hawks (and perhaps even an owl or two), and then finds the relevant "field marks".

The dexterity with which a birdwatcher negotiates the pages of the text has little to do with physiological co-ordination, but much more to do with taxonomy. One of the quickest routes to the appropriate picture is to look up the bird's name in the index, but, to take this route, the birder must already know or be able to guess the species name. Other less precise indexes are based, depending on the field guide, on the general silhouette, "family" resemblance of the bird, dominant colour, and type of habitat.

4.2.3 Troubles with specimens "not in the book"

Birdwatchers not uncommonly sight specimens which they cannot "find" in the field guide. Even after obtaining a "good look" at the bird in the field, repeated searching in the book reveals no description or illustration which appears to do the job. There are many possible reasons for this, but the beginning birdwatcher almost never concludes that she has discovered an uncommon species or that the book had omitted a common species. Rather, the novice typically accepts the authority of the text while attributing the trouble to her inexperience, problems in perspective, or to an atypical appearance of the particular individual or local variant of the species.

Although this list of "troubles" barely begins to specify the difficulties encountered while playing the literary language game, we do not mean to imply that all field identifications are beset by troubles. Some birds in the field are "seen at a glance". Specimen and species identity are appropriated within the same instant. How this is accomplished is complex and worthy of a more extended discussion than we can give it here. We will note only that "seeing at a glance" is circumstantial, dependent upon local expertise, and is not simply dependent on intrinsic "markings" which make some species very different from all others which share their range. We have focussed on instances of "troubled" identifications because they extend the "moment" of identification and reveal effortful attempts to bring naturalistic observation within the province of textual order.

5. Conclusion

In this paper we have considered the organization of textual materials – notably lists and field manuals – used in amateur birdwatching. We have focused on the place of these texts in a particular form of taxonomic application which we have called a "literary language game", and we have argued that the texts provide a descriptive organization to the craft of *seeing* species in the field. Thus, instead of relying upon perceptual metaphors for elucidating the theoretic organization of observation, we have preferred the metaphor of "reading" and "writing" with the naturalistic gaze.

Birdwatcher's lists, we have tried to show, are not just columns of names (although in further work it would be interesting to examine the concrete features of particular lists). They are accountable reports of observations. This is especially so when the lists are compiled as part of socially organized practice (and even the lists of the solitary birdwatcher are part of such a practice, though perhaps freed from some constraints that cover collaborative field surveys). The accountability of the lists pervades the methodical "look". "Getting a good look" is clearly a list-organized phenomenon.

The activity of birdwatching is also an exemplar for studying the place of an instructional text in order of practice. This is an established topic in ethnomethodology, which has been applied in analyses of work in scientific and other fields (Garfinkel, 1967; Lynch et al., 1983; Suchman, 1987, Amerine and Bilmes, 1988). Here we have described some of the similarities and differences between field guides, and some of the troubles that arise from using three texts in the field. This has allowed us to identify some of the contingencies arising within the hermeneutic circle comprised of perceptive "readings" in a wordly field in reference to an authoritative text.

For reasons of space, and because we have restricted our focus to the place of texts in field observation, we have not dealt with a number of other themes. In further work we intend to develop the issue of "getting a good look" beyond the brief mentions that we have made of it here, and we will devote more explicit attention to several other topics on the embodied and social interactional accomplishment of field observation. These include the location of birdwatching sites and "species places", the various forms of intervention in the field for exposing or eliciting birds, the use of optical technology and its associated problems, and the interactional and institutional organization of group birdwatching.

It is also important to explore the applicability of our exemplar to scientific practice. Although we find birdwatching interesting and enjoyable, we do not aim to pursue a "sociology of birdwatching" for its own sake. The question then is, what does our account of birdwatching have to do with practices in science? Our answer is tentative. Arguably the apprenticeship of a novice to the "normal science" of birdwatching reflects more general processes of apprenticeship in which textually ordered "knowledge" is elaborated in the course of practical investigation.[15] If this is the case, then the study of birdwatching (and other "trouble-prone" lay observational activities) speaks directly to a burgeoning interest in the social studies of science about the relationship between textual order and scientific practice.[16] Writing and documentation are seen as pervading laboratory practice, as lending organization, packaging and stabilizing sense, and as constitutive to the process of building and circulating claims. Literary artifacts (O'Neill, 1980) absorb and articulate orders of things within orders of discourse. So it is with texts in birdwatching. The strategies of list-building and field-guide design and the constitutive troubles that these bring in their wake – strategies that are relatively transparent to the outsider – may find their analogues deep within esoteric scientific practice where they are correspondingly more difficult to detect.

Whether the analogy that we have drawn between birdwatching

and science is defensible is in part a function of the uses to which it is put and of the assumptions one makes about the nature of science – its unity or heterogeneity, and its relationship to the "ordinary" institutions of language and practice. If, however, the analogy is permissible (and we note that Kuhn, 1977a:390ff., for one, has made use of birdwatching metaphors in his writing), then our argument does three things. First, it supports those who claim that the "scientific mind" may be treated as a social construction rather than something which is located within the hardware of the brain.[17] Second, it suggests that there is much to be learned by considering the way in which rather mundane "ecological" practices are managed in scientific work – such as matters of the spatial and sequential distribution of instruments, texts and social relations. And third, and more modestly, it points to the methodological utility of using "homely" examples in order to obtain purchase on esoteric and technical practices.

Notes

1. While we will not follow up many of the nuances of this term, we find Wittgenstein's (1953:214) discussion very much to the point:

   > Aspect-blindness will be *akin* to the lack of a "musical ear".
   >
   > The importance of this concept lies in the connexion between the concepts of 'seeing an aspect' and 'experiencing the meaning of a word'. For we want to ask 'What would you be missing if you did not *experience* the meaning of a word?'

2. Here we are borrowing and no doubt misusing Wittgenstein's (1953: § 7) terminology to speak of a practical use of words in an apprenticeship to a language:

   > ... one party calls out the words and the other acts on them. In instruction in the language the following process will occur: the learner *names* the objects; that is, he utters a word when the teacher points to the stone.

   In the *literary* language game, one of the "parties" to the game is a field guide, and a list acts as a textual expression within the socially organized competency of birdwatching. The literary language game is, of course, an ideal-typical construction; although one based upon our own practical experience and our observations of other novices.
3. See Kastner (1986) for a history of competing classificatory schemes in birdwatching, and Farber (1982) for an analogous account in ornithology.

4. This point was driven home to one of the authors, an American, while travelling in Britain. He purchased an inexpensive field guide to British birds and used it as a reference while travelling through parts to Britain. The field guide proved to be quite easy to use, and he accumulated a short list of species seen in backyards and along highways and pathways. Later, when shown another field guide with a much more extensive catalogue, he was disappointed to find that many of the species he had "counted" while using the simple manual could no longer be distinguished from cognate species included in the more complete manual.
5. Using Myers' (1988) scheme, we can arrange the conventions used in each field guide along a continuum from "abstract" (Peterson's schematic paintings) through "realistic" (the Audubon guide's photographs), with the "naturalistic" diorama (National Geographic guide) being intermediate. Placement of an illustration along the continuum is demonstrated by its inclusion or exclusion of what Myers calls "gratuitous" detail (aspects of the picture having no direct bearing on its analytic use in the text). Like Myers, we are concerned with the way in which such conventions provide a text's claims with variable analytic and realistic authority. However, the illustrative conventions used in field guides, unlike those in the text Myers analyzes, are primarily relevant as identification instruments. The pictures are operated as "ways of seeing" within the literary language game of birdwatching, and not as polemical moments in a text's argument.
6. Plate 44 in Peterson (1947) and pages 210–213 in Peterson (1980).
7. See Barnes' (1977:7–9) discussion of pictoral representation. He writes of a figure depicting the muscles of the arm from an anatomy textbook that

   It is designed to facilitate recognition and naming of an esoteric activity. *Therefore*, it is not a rendering of a particular arm. Despite being apparently realistic it is intentionally a schemata. (Barnes, 1977:7, his italics)

8. Compare the tail shapes in the photographs in the Audubon guide.
9. This point is discussed in Law and Lodge, 1984:94.
10. Myers' (1988) uses the analogy of museum diorama to describe a typical style of drawing where realist depiction, using some conventions of photography, is combined with non-photographic synthesis of paradigmatic views, surroundings, and apparent "activities".
11. The following troubles are described in typified fashion, sometimes exemplified by recollected incidents. By describing these in typified form we are claiming that they recur, that they have been and will be experienced as regular features of the novice's language game. We are not, of course, wanting to suggest that our account represents a full description of that language game.
12. In addition, the Audubon guide suffers a further disadvantage which it shares with the earlier though not the more recent Peterson edition – that its illustrations and text are located on different pages. Since it is

difficult to tell what features of similarity and difference should be attended to by consulting the photographs alone, in practice this means that the birdwatcher has to move rapidly between two parts of the guide and the binoculars in order to achieve many identifications – a task of needless complexity, given the availability of simpler field manuals.

13. Compare the photographs of the tails in the Audubon guide!
14. The detection of markings through binoculars is itself a rich topic which we will consider in another paper.
15. There is, of course, no single analogy. There are numerous kinds of birdwatching, just as there are innumerable forms of observational activity in science. There is also an open variety of themes: intervention, use of equipment, social interaction and authoritative description. In the present paper we have considered only a very limited part of the practice of birdwatching.
16. See, inter alia, Latour and Woolgar (1979); Woolgar (1980); Knorr-Cetina (1981); Bazerman (1981); Law and Williams (1982); Callon (1986); Callon, Law and Rip (1986); Griesemer and Star (1986); Callon and Law (1987); Morrison (1988); Yearley (1988); Gilbert and Mulkay (1984); Lynch (1985a; 1985b; 1988); Latour (1986); Law (1986a); Myers (1988); and Amann and Knorr-Cetina (1988).
17. See Star (1987) for details of the social construction of localisation theories.

References

Alpers, S. (1983). *The art of describing: Dutch art in the seventeenth century*. Chicago: University of Chicago Press.

Amann, K., and Knorr-Cetina, K.D. (1988). The fixation of visual evidence. *Human Studies* 11(2–3):133–169.

Amerine, R., and Bilmes, J. (1988). Following instructions. *Human Studies.*

Barnes, B. (1977). *Interests and the growth of knowledge*. London: Routledge.

Bazerman, C. (1981). What written knowledge does: Three examples of academic discourse. *Philosophy of the Social Sciences* 11(3):361–387.

Bernstein, C. (1984). Details on details: Describing a bird. *Western Tanager* 50(6):1–3.

Callon, M. (1986). Some elements of a sociology of translation: Domestication of the scallops and fishermen of St. Brieuc Bay. In J. Law (Ed.), *Power, action and belief: A new sociology of knowledge*?, 196–233. Sociological Review Monograph 32, London: Routledge and Kegan Paul.

Callon, M., and Law, J. (1987). Economic markets and scientific innovation: Notes on the construction of sociotechnical networks. *Knowledge and Society*. Forthcoming.

Callon, M., Law, J., and Rip. A., Eds. (1986). *Mapping the dynamics of science and technology: The sociology of science in the real world.* London: Macmillan.

Farber, P. (1982). *The emergence of orthnithology as a scientific discipline: 1760–1850.* Dordrecht, Boston and London: Reidel.

Garfinkel, H. (1967). *Studies in Ethnomethodology.* Englewood Cliffs: Prentice Hall.

Garrett, K.L. (1986a). Field tips: The single field mark syndrome. *Western Tanager* 52(5):1–2.

Garrett, K.L. (1986b). Field tips: The Bishop test. *Western Tanager* 52(10): 1–3.

Gilbert, G.N., and Mulkay, M.J. (1984). *Opening Pandora's box: A sociological analysis of scientists' discourse.* Cambridge: Cambridge University Press.

Gombrich, E.H. (1960). *Art and illusion: A study in the psychology of pictorial representation.* Princeton: Princeton University Press.

Griesemer, J., and Star, S.L. (1986). Paper presented to the Society for Social Studies of Science Annual Meeting, Pittsburgh, PA.

Helm, D., et al. (1987). *The interactional order.* New York: Irvington.

Hosking, E., with MacDonnell, K. (1979). *Eric Hosking's birds: Fifty years of photographing wildlife.* London: Pelham Books.

Kastner, J. (1986). *A world of watchers.* New York: Knopf.

Knorr-Cetina, K.D. (1981). *The manufacture of knowledge: An essay on the constructivist and contextual nature of science.* Oxford: Pergamon Press.

Knorr, K.D., Krohn, R., and Whitley, R.D. Eds. (1980). *The social process of scientific investigation. Sociology of the sciences* 4. Dordrecht, Boston and London: Reidel.

Knorr-Cetina, K.D., and Mulkay, M.J., Eds. (1983). *Science observed: Perspectives on the social study of science.* London and Beverly Hills: Sage.

Kuhn, T.S. (1977a). Second thoughts on paradigms. In T.S. Kuhn (Ed.), *The essential tension: Selected studies in scientific tradition and change,* 293–319. Chicago and London: University of Chicago Press.

Kuhn, T.S. (1977b). *The essential tension: Selected studies in scientific tradition and change.* Chicago and London: University of Chicago Press.

Latour, B. (1986). Visualisation and cognition. *Knowledge and Society* 6: 1–40.

Latour, B., and Woolgar, S. (1979). *Laboratory life: The social construction of scientific facts.* London and Beverly Hills: Sage.

Law, J. (1986a). On power and its tactics: A view from the sociology of science. *Sociological Review* 34:1–37.

Law, J., Ed. (1986b). *Power, action and belief: A new sociology of knowledge?* Sociological Review Monograph 32. London: Routledge and Kegan Paul.

Law, J., and Lodge, P. (1984). *Science for social scientists.* London: Macmillan.

Law, J., and Williams, R. (1982). Putting facts together: A study of scientific persuasion. *Social Studies of Science* 12(4):535–558.

Lynch, M. (1985a). *Art and artifact in laboratory science: A study of shop work and shop talk in a research laboratory*. London: Routledge and Kegan Paul.

Lynch, M. (1985b). Discipline and the material form of images: An analysis of scientific visibility. *Social Studies of Science* 15:37–66.

Lynch, M. (1988). The externalized retina: Selection and mathematization in the visual documentation of objects in the life sciences. *Human Studies* 11(2–3):201–234.

Lynch, M., Livingston, E., and Garfinkel, H. (1983). Temporal order in laboratory work, 205–238. In K.D. Knorr-Cetina and M.J. Mulkay (Eds.), *Science observed: Perspectives on the social study of science*, 205–238. London and Beverley Hills: Sage.

Morrison, K. (1988). Some researchable recurrences in science and situated science inquiry. In D. Helm et al. (Eds.), *The interactional order*. New York: Irvington.

Myers, G. (1988). Every picture tells a story: Illustrations in E.O. Wilson's *Sociobiology*. *Human Studies* 11(2–3):235–269.

National Geographic Society (1983). *A field guide to the birds of North America*. Washington, DC: National Geographic Society.

O'Neill, J. (1980). The literary production of natural and social science inquiry. *Canadian Journal of Sociology* 6:105–120.

Peterson, R.T. (1934; 1939; 1947; 1980). *A field guide to the birds of the Eastern United States*. Boston: Houghton Mifflin.

Peterson, R.T. (1941; 1961). *A field guide to Western Birds*. Boston: Houghton Mifflin.

Star, S.L. (1988). *Regions of the mind: British brain research, 1870–1906*. Palo Alto: Stanford University Press.

Suchman, L. (1987). *Plans and situated actions: The problem of machine-human communication*. Cambridge: Cambridge University Press.

Udvardy, M.D.F. (1977). *The Audubon Society field guide to the North American Birds, Western region*. New York: Knopf.

Wittgenstein, L. (1953). *Philosophical Investigations*, trans. G.E.M. Anscombe. New York: Macmillan.

Woolgar, S. (1980). Discovery: Logic and sequence in a scientific text. In K.D. Knorr, R. Krohn and R.D. Whitley (Eds.), *The social process of scientific investigation. Sociology of the Sciences* 4, 239–268. Dorrecht, Boston and London: D. Reidel.

Yearley, S. (1988). The dictates of method and policy: Interpretational structures in the representation of scientific work. *Human Studies* 11 (2–3):341–359.

Representing practice in cognitive science*

LUCY A. SUCHMAN
System Sciences Laboratory, Xerox Corporation, 3333 Coyote Hill Road, Palo Alto, CA 94304, USA

1. Introduction

Recent social studies of science take as a central concern the relationship between various representational devices and scientific practice (see Tibbett, 1988, and Lynch and Woolgar, 1988, in this special issue.) Representational devices include models, diagrams, formulae, records, traces and a host of other artifacts taken to stand for the structure of an investigated phenomenon. Several premises underlie the study of the relation of such devices to scientific practice. First, that it is through these devices that the regularity, reproducibility and objectivity both of phenomena and of the methods by which they are found are established. Second, that representational devices have a systematic but necessarily contingent and *ad hoc* relation to scientific practices. And third, that representational technologies are central to how scientific work gets done. To date science studies have concentrated on the physical and biological sciences (see for example Collins, 1985; Latour and Woolgar, 1979; Garfinkel et al., 1981; Knorr-Cetina, 1981; Lynch, 1985; Lynch et al., 1983). This paper joins with others (Woolgar, 1985; Collins, 1987) in directing attention to a new arena of scientific practice; namely, cognitive science.

* This paper and the work that it reports have benefited substantially from discussions with my collaborators Randy Trigg and Brigitte Jordan. For developing observations on the use of whiteboards I am indebted to Randy Trigg, John Tang and to members of the Interaction Analysis Group at Xerox PARC; Christina Allen, Stephanie Behrend, Sara Bly, Tom Finholt, George Goodman, Austin Henderson, Brigitte Jordan, Jane Laursen, Susan Newman, Janice Singer, and Debbie Tatar.

In turning to cognitive science as a subject of sociological inquiry we are faced with an outstanding issue concerning the relation of representation to practice. The issue can be formulated, at least initially, as follows: Ethnomethodological studies of the physical and biological sciences eschew any interest in the adequacy of scientific representations as other than a members' concern. The point of such studies is specifically *not* to find ironies in the relation between analysts' constructions of the phenomenon and those of practitioners (Garfinkel, 1967:viii; Woolgar, 1983.) Rather, the analyst's task is to see how it is that practitioners come to whatever understanding of the phenomenon they come to as the identifying accomplishment of their scientific practice. In turning to cognitive science, however, one turns to a science whose phenomenon of interest itself *is* practice. For cognitive science theorizing, the object is mind and its manifestation in rational action. And in designing so-called intelligent computer systems, representations of practice – expert/novice instruction, medical diagnosis, electronic troubleshooting and the like – provide the grounds for achieving rationality in the behavior of the machine.

In this paper I consider two distinct but related conceptions of the notion of "representing practice" with respect to cognitive science, through a discussion of two studies. The first study, recently completed, looks at the ways in which cognitive scientists depict the nature and operation of social practice, as part of their own agenda for the design of intelligent machines. These ways include the representation of practice as logical relations between conditions and actions, and the design of artifacts that embody such representations. The second study, just underway, looks at the representational practices of cognitive scientists, through a detailed analysis of researchers engaged in collaborative design work at a "whiteboard."[1] Together these studies consider representing practice as both the object of cognitive scientists' work and as sociology's subject matter.

2. Artificial intelligence and interactional competence

The term "cognitive science" came into use in the 1970s to refer to a convergence of interest over the preceding 20 years among neurophysiologists, psychologists, linguists, cognitive anthropologists, and later computer scientists, in the possibility of an integrated science of cognition (for an enthusiastic history see Gardner, 1985). The commitment both to cognition and to science was, at least initially, an important part of the story. At the turn of the century, the recognized method for studying human mental life was introspection and, insofar as introspection was not amenable to the emerging canons of scientific method, the study of cognition seemed doomed to be irremediably unscientific. In reaction to that prospect, the behaviorists posited that human action should be investigated in terms of publicly observable, mechanistically describable relations between the organism and its environment. In the name of turning cognitive studies into a science, the study of cognition as the study of something apart from conditioned behavior was effectively abandoned in mainstream psychology.

Cognitive science, in this respect, was a project to bring thought, or meaning, back into the study of human action while preserving the commitment to scientism. Cognitive science reclaims mentalist constructs like beliefs, desires, intentions, planning and problem-solving. Once again human purposes are the basis for cognitive psychology, but this time without the unconstrained speculation of the introspectionists. The study of cognition is to be empiricized not by a strict adherence to behaviorism, but by the use of a new technology; namely, the computer.

The branch of cognitive science most dedicated to the computer is Artificial Intelligence. The sub-field of AI arose as advances in computing technology were tied to developments in neurophysiological and mathematical theories of information. The requirements of computer modeling, of an "information processing psychology," seem both to make theoretical sense and to provide the accountability that will make it possible to pursue a science of otherwise inaccessible mental phenomena. If underlying mental processes can be modelled on the computer so as to produce the right outward behavior, the argument goes, the model can be viewed as having passed at least a sufficiency test of its psychological validity.

A leading idea in cognitive science is that mind is best viewed as neither substance nor as insubstantial, but as an abstractable structure implementable in any number of possible physical substrates. Intelligence, on this view, is only incidentally embodied in the neurophysiology of the human brain. What is essential about intelligence can be abstracted from that particular, albeit highly successful substrate and embodied in an unknown range of alternative forms. The commitment to an abstract, disembodied account of cognition, on the one hand, and to an account of cognition that can be physically embodied in a computer, on the other, has led to a view of intelligence that takes it to be first and foremost mental operations and only secondarily, and as an epiphenomenon, the "execution" of situated actions.

While intelligence is taken by cognitive science, without much question, to be a faculty of individual minds, the measure of success for the AI project is and must be an essentially social one. Evidence for intelligence, after all, is just the observable rationality of the machine's output relative to its input. This sociological basis for machine intelligence is implicit in the so-called Turing Test, by now more an object of cognitive science folklore than a part of working practice. Turing (1950) argued that if a machine could be made to respond to questions in such a way that a person asking the questions could not distinguish between the machine and another human being, the machine would have to be described as intelligent.[2] Turing expressly dismissed as a possible objection to his proposed test that, although the machine might succeed in the game, it could succeed through means that bear no resemblance to human thought. Turing's contention was precisely that success at performing the game, regardless of mechanism, is sufficient evidence for intelligence (1950:435). The Turing test thereby became the canonical form of the argument that if two information-processors, subject to the same input stimuli, produce indistinguishable output behavior, then regardless of the identity of their internal operations one processor is essentially equivalent to the other.

The lines of controversy raised by the Turing test were drawn over a family of programs developed by Joseph Weizenbaum in the 1960s under the name ELIZA, designed to support "natural language conversation" with a computer (Weizenbaum, 1983).

Of the name ELIZA, Weizenbaum writes:

> Its name was chosen to emphasize that it may be incrementally improved by its users, since its language abilities may be continually improved by a "teacher." Like the Eliza of *Pygmalion* fame, it can be made to appear even more civilized, the relation of appearance to reality, however, remaining in the domain of the playwright. (p. 23)

Anecdotal reports of occasions on which people, approaching the teletype to one of the ELIZA programs and believing it to be connected to a colleague, engaged in some amount of "interaction" without detecting the true nature of their respondent led many to assert that Weizenbaum's program had passed a simple form of the Turing test. Weizenbaum himself, however, denied the intelligence of the program on the basis of the underlying mechanism, which he described as "a mere collection of procedures" (p. 23):

> The gross procedure of the program is quite simple; the text [written by the human participant] is read and inspected for the presence of a *keyword*. If such a word is found, the sentence is transformed according to a *rule* associated with the keyword, if not a content-free remark or, under certain conditions, an earlier transformation is retrieved. The text so computed or retrieved is then printed out. (p. 24, original emphasis)

The design of the ELIZA programs exploits the natural inclination of people to make use of the "documentary method of interpretation" (see Garfinkel, 1967:Ch. 3): to take appearances as evidence for, or the document of an ascribed underlying reality while taking the reality so ascribed as a resource for the interpretation of the appearance. In a contrived situation that, though designed independently and not with them in mind, closely parallels both the "Turing test" and encounters with Weizenbaum's ELIZA programs, Garfinkel set out to test the documentatry method in the context of counseling. Students were asked to direct questions concerning their personal problems to someone they knew to be a student counselor, seated in another room. They were restricted to questions that could take yes/no answers, and those answers were given by the counselor on a random basis. For the students, the counselor's answers were motivated by the questions. That is to say, by taking each answer as evidence for what the counselor "had in mind," the students were able to find a deliberate pattern in the exchange that explicated the significance and relevance of each new response as an answer to their question:

> The underlying pattern was elaborated and compounded over the series of exchanges and was accommodated to each present "answer" so as to maintain the "course of advice," to elaborate what had "really been advised" previously, and to motivate the new possibilities as emerging features of the problem. (1967:90)

The ELIZA programs and Garfinkel's counselor experiment demonstrate the generality of the documentary method and the extent to which the meaning of actions is constituted not by actors' intentions but through the interpretive activity of recipients. Users of ELIZA and Garfinkel's students are able to construct out of the mechanical "responses" of the former and the random "responses" of the latter a response to *their* questions. This clearly poses a problem for Turing test criteria of intelligence and for the test of intentionality proposed by Dennett (1978:Ch. 1), who argues that intentional systems are just those whose behavior is conveniently made sense of in intentional terms. ELIZA and the counselor clearly meet that criterion, yet they show at the same time the inadequacy of that measure for intentional interaction. The injunction for the counselor is precisely that he or she *not* interact. The counselor's "responses" are not responses to the student's questions, nor are the interpretations that the student offers subject to any remediation of misunderstanding by the counselor. Or rather, there *is no* notion of misunderstanding, insofar as in the absence of the counselor's point of view any understanding on the part of the student that "works" will do. In human communication in contrast there are two "students," *both* engaged in making sense out of the actions of the other, in making their own actions sensible, in assessing the senses made, and in looking for evidence of misunderstanding. It is just this highly contingent and reciprocal process that we call "interaction."

For behavior to be not only intelligible but intentional, it seems, there must be something about the *actor* that gives her action its senses. As participants in interaction we see our work not as the single-handed construction of meaning but as a kind of reading off from the action of the actor's underlying intent. This common sense view is adopted by cognitive scientists, who take actions to reflect the underlying cognitive mechanism or plans that generate them. The representation of those mechanisms or plans, on this view, is effectively the representation of practice.

3. Plans as determinants of action

The identification of intent with a plan-for-action is explicit in the writing of philosophers of action supportive of artificial intelligence research like Margaret Boden (1973) who writes:

> unless an intention is thought of as an action-plan that can draw upon background knowledge and utilize it in the guidance of behavior one cannot understand how intentions function in real life. (pp. 27–28)

A logical extension of Boden's view, particularly given an interest in rendering it more computable, is the view that plans actually are prescriptions or instructions for action. Traditional sociology similarly posits an instrumentally rational actor whose choice among alternative means to a given end is mediated by norms of behavior that the culture provides – an actor Garfinkel dubbs the "cultural dope":

> By "cultural dope" I refer to the man-in-the-sociologist's-society who produces the stable features of the society by acting in compliance with preestablished and legitimate alternatives of action that the common culture provides. (1967: 68)

Cognitive science embraces this normative view of action in the form of the *planning model.* The model assumes that in acting purposefully actors are constructing and executing plans, condition/action rules, or some other form of representation that controls, and therefore must be prerequisite to, actions-in-the-world. An early and seminal articulation of this view came from Miller, Galanter and Pribram, in *Plans and the Structure of Behavior* (1960):

> Any complete description of behavior should be adequate to serve as a set of instructions, that is, it should have the characteristics of a plan that could guide the action described. When we speak of a plan ... the term will refer to a hierarchy of instructions ... *A plan is any hierarchical process in the organism that can control the order in which a sequence of operations is to be performed.*
>
> A Plan is, for an organism, essentially the same as a program for a computer ... we regard a computer program that simulates certain features of an organism's behavior as a theory about the organismic Plan that generated the behavior.
>
> Moreover, we shall also use the term "Plan" to designate a rough sketch of some course of action ... as well as the completely detailed specification of every detailed operation ... We shall say that a creature is executing a particular Plan when in fact that Plan is controlling the sequence of operations he is carrying out. (p. 17, original emphasis)

With Miller et al., the view that purposeful action is planned assumes the status of a psychological theory compatible with the interest in a mechanistic, computationally tractable account of intelligent action. The identification of intentions with plans, and plans with programs, leads to an identification of representation and action that supports the notion of "designing" intelligent actors. Once representations are taken to control human actions, the possibility of devising formalisms that could specify the actions of "artificial agents" becomes plausible. Actions are described by preconditions, that is, what must be true to enable the action, and postconditions, what must be true after the action has occurred. By improving upon or completing our common sense notions of the structure of action, the structure is now represented not only as an empirically ascertained set of behavioral patterns or a plausible sequence of actions but as an hierarchical plan. The plan reduces, moreover, to a detailed set of instructions that actually serves as the program that controls the action. At this point, the plan as stipulated becomes substitutable for the action, insofar as the action is viewed as derived from the plan. And once this substitution is done, the theory is self-sustaining: the problem of action is *assumed* to be solved by the planning model, and the task that remains is to refine the model.

4. Plans as resources for action

Taken as the determinants of what people do, plans provide both a device by which practice can be represented in cognitive science and a solution to the problem of purposeful action. If we apply an ethnomethodological inversion[3] to the cognitive science view, however, plans take on a different status. Rather than describing the mechanism by which action is generated and a solution to the analysts' problem, plans are common sense constructs produced and used by actors engaged in everyday practice. As such, they are not the solution to the problem of practice but part of the subject matter. While plans provide useful ways of talking and reasoning about action, their relation to the action's production is an open question.

One can see clearly the descriptive or interpretive function of talk about intentions, and its problematic relation to production, in the case of our talk about babies.[4] Nursing babies are very good at finding milk. If you touch a baby on the cheek, it will move its head in the direction of the touch. Similarly, if you put your finger on the baby's lips, it will suck. In some sense we would say, in describing the baby's behavior, that the baby "knows how to get food." Yet to suggest that the baby "has a goal" of finding food in the form of a representation of the actions involved, or performs computations on data structures that include the string "milk" to reach that goal, seems somehow implausible. It is not that all behavior can be reduced to the kind of reflex action of a nursing baby, or that some behavior is not importantly symbolic. The point is that the intentional description, however useful, doesn't distinguish those things.

At the same time, such description is clearly a resource. Our imagined projections and retrospective reconstructions are the principal means by which we catch hold of situated action and reason about it, while situated action itself is essentially transparent to us as actors. In contemplating the descent of a problematic series of rapids in a canoe, for example, one is very likely to sit for a while above the falls and plan one's descent.[5] So one might think something like "I'll get as far over to the left as possible, try to make it between those two large rocks, then backferry hard to the right to make it around that next bunch." A great deal of deliberation, discussion, simulation, and reconstruction may go into such a plan and to the construction of alternate plans as well. But in no case – and this is the crucial point – do such plans control action in any strict sense of the word "control." Whatever their number or the range of their contingency, plans stop short of the actual business of getting you through the falls. When it really comes down to the details of getting the actions done, *in situ*, you rely not on the plan but on whatever embodied skills of handling a canoe, responding to currents and the like are available to you. The purpose of the plan, in other words, is not literally to get you through the rapids, but rather to position you in such a way that you have the best possible conditions under which to use those embodied skills on which, in the final analysis, your success depends.

The planning model takes off from our common sense preoccupation with the anticipation of action and the review of its outcomes and attempts to systematize that reasoning as a model for situated practice itself. These examples, however, suggest an alternative view of the relationship between plans, as representations of conditions and actions, and situated practice. Situated practice comprises moment-by-moment interactions with our environment more and less informed by reference to representations of conditions and of actions, and more and less available to representation themselves. The function of planning is not to provide a specification or control structure for such local interactions, but rather to orient us in a way that will allow us, through the local interactions, to respond to some contingencies of our environment and to avoid others. As Agre and Chapman put it "[m]ost of the work of using a plan is in determining its relevance to the successive concrete situations that occur during the activity it helps to organize" (1987a).[6] Plans specify actions just to the level that specification is useful; they are vague with respect to the details of action precisely at the level at which it makes sense to forego specification and rely on the availability of a contingent and necessarily *ad hoc* response. Plans are not the determinants of action, in sum, but rather are resources to be constructed and consulted by actors before and after the fact.

5. Engineering interaction

Adherents of the planning model in AI view interaction just as an extension of the planning problem from a single individual to two or more individuals acting in concert. In a 1983 paper on recognizing intentions, James Allen puts it this way:

> Let us start with an intuitive description of what we think occurs when one agent A asks a question of another agent B which B then answers. A has some goal; s/he creates a plan (plan construction) that involves asking B a question whose answer will provide some information needed in order to achieve the goal. A then executes this plan, asking B the question. B interprets the question, and attempts to infer A's plan (plan inference). (p. 110)

The problem for interaction, on this view, is to recognize the actions of others as the expression of their underlying plans. The appropriateness of a response turns on that analysis, from which, in turn the hearer then adopts new goals and plans her own utterances to achieve them. On this model, Searle's speech act theory seems to offer some initial guidelines for computational models of communication. Searle's conditions of satisfaction for the successful performance of speech acts are read as the speech act's "preconditions," while its illocutionary force is the desired "effect:"

> Utterances are produced by actions (speech acts) that are executed in order to have some effect on the hearer. This effect typically involves modifying the hearer's beliefs or goals. A speech act, like any other action, may be observed by the hearer and may allow the hearer to infer what the speaker's plan is. (Allen, 1983:108)

Given this view, the design of interactive computer systems affords a kind of natural laboratory in which to see what happens when artifacts embodying the planning model of action encounter people engaged in situated activity. The practical problem with which the designer of an interactive machine must contend is how to ensure that the machine responds appropriately to the user's actions. The design strategy for plan-based systems is essentially to *specify* an appropriate linkage between user actions and machine states. This strategy assumes that the behavior of both user and machine can be represented in advance as a plan that not only projects but determines their local interaction.

A conversation analysis of such encounters, however, reveals that while interaction between people and machines requires essentially the same interpretive work that characterizes interaction between people, fundamentally different resources are available to the "participants" (for a full account see Suchman, 1987). In particular, people make use of a rich array of experience, embodied skill, material evidence, communicative competence and members' knowledge in finding the intelligibility of actions and events, in making their own actions sensible, and in managing the troubles in understanding that inevitably arise. Due to constraints on the machine's access to the situation of the user's inquiry, however, breaches in understanding that for face-to-face interaction would be trivial in terms of detection and repair become "fatal" for human-machine communication (see Jordan and Fuller, 1974). The result is an asymmetry that severely limits the scope of interaction between people and machines.

Because of this asymmetry, engineering human-machine interaction becomes less a matter of simulating human communication than of finding alternatives to interaction's situated properties. Those properties and the subtlety of their operation are nicely illustrated in the following fragment of naturally occurring conversation:

A: Are you going to be here for ten minutes?
B: Go ahead and take your break. Take longer if you want.
A: I'll just be outside on the porch. Call me if you need me.
B: OK. Don't worry.
(Gumperz, 1982:326)

In his analysis of this fragment Gumperz points out that B's response to A's question clearly indicates that B interprets the question as an indirect request that B stay in the office while A takes a break, and by her reply A confirms that interpretation. B's interpretation accords with a categorization of A's question as an indirect speech act (Searle, 1979), and with Grice's discussion of implicature (1975); that is, B assumes that A is cooperating, and that her question must be relevant, therefore B searches her mind for some possible context or interpretive frame that would make sense of the question, and comes up with the break. But, Gumperz points out, this analysis begs the question of how B arrives at the right inference:

> What is it about the situation that leads her to think A is talking about taking a break? A common sociolinguistic procedure in such cases is to attempt to formulate discourse rules such as the following: "If a secretary in an office around break time asks a co-worker a question seeking information about the co-worker's plans for the period usually allotted for breaks, interpret it as a request to take her break." Such rules are difficult to formulate and in any case are neither sufficiently general to cover a wide enough range of situations nor specific enough to predict responses. An alternative approach is to consider the pragmatics of questioning and to argue that questioning is semantically related to requesting, and that there are a number of contexts in which questions can be interpreted as requests. While such semantic processes clearly channel conversational inference, there is nothing in this type of explanation that refers to taking a break. (1982:326)

The problem that Gumperz identifies here applies equally to attempts to account for inferences such as B's by arguing that she "recognizes" A's plan to take a break. Clearly she does: the outstanding question is how. While we can always construct a *post hoc* account that explains interpretation in terms of knowledge of typical situations and motives, it remains the case that neither typifications of intent nor general rules for its expression are sufficient to account for the mutual intelligibility of our situated action. In the final analysis, attempts to represent intentions and rules for their recognition seem to beg the question of situated interpretation, rather than answering it.

6. Cognitive science's situated practice

The decontextualized models of action embraced by the majority of cognitive science researchers stand in contrast to the situated structuring of their own scientific practice.[7] Our current research examines how various "inscription devices" (Latour and Woolgar, 1979) or technologies for representation are used by cognitive scientists and systems designers engaged in the collaborative invention of new computational artifacts. A common technology for representation in our laboratory is the "whiteboard." We begin with the observation, due to Livingston (1978), that the inscriptions on a whiteboard – lists, sketches, lines of code, lines of text and the like – are produced through activities that are not themselves reconstructable from these "docile records" (Garfinkel and Burns, 1979). Methodologically, this means that the core of our data must be audiovisual recordings of the moment-by-moment interactions through which the inscriptions are produced. Made observable, the organization of activities that produce marks on the whiteboard and give them their significance, and the function of marks in the structure of the activity, become our research problem.

Our starting assumption is that the use of the whiteboard both supports and is organized by the structure of face-to-face interaction. On that assumption, our analysis is aimed at uncovering the relationship between (i) the organization of face-to-face interaction, (ii) the collaborative production of the work at hand and (iii) the use of the whiteboard as an interactional and representational resource. From the video corpus we aim to identify systematic practices of whiteboard use, with a focus on just how those practices and the inscriptions they produce constitute resources for particular occasions of technical work. Some initial conjectures are the following:[8]

6.1 *The whiteboard is a medium for the construction of concrete conceptual objects*

Inscriptions on the whiteboard are conceptual in that they stand for phenomena that are figurative, hypothetical, imagined, proposed or otherwise not immediately present, but they are also concrete – visible, tangible marks that can be pointed to, modified, erased and reproduced. Over the work's course topics of talk are visibly constituted on the board, becoming items to be considered, revised, adopted and reconsidered. Technical objects once represented can be "run," subject to various scenarios, examined for their structure and so on. Conceptual objects rendered concrete, in sum, become available for development and change.

6.2 *The whiteboard structures mutual orientation to a shared interactional space*

Through their orientation in seating arrangements, body positions, gesture and talk, collaborators turn the whiteboard and its marks into objects in a shared space. We see designers, on first sitting down to work, "referring" in their talk and gestures to a whiteboard on which nothing has yet been written. Mutual engagement is demonstrated (or not) by attention either to the other(s) or to the shared space of the board. Bodily movements of, for example, standing at the board with marker raised or stepping back with folded arms display the status of objects as incomplete, problematical, satisfactory and the like.

6.3 *Talk and writing are systematically organized*

Skilled work at the whiteboard effectively exploits the "simplest systematics for the organization of turn-taking for conversation" (see Sacks et al., 1974) in the sequential organization of turns at talk and writing. The board provides a second interactional floor, co-extensive and sequentially interleaved with that of talk. So, for example, the board may be used in taking and holding the floor, or in maintaining some writing activity while passing up a turn at talk. Writing done during another's talk (may (a) document the talk and thereby display the writer's understanding, (b) continue the writer's previous turn or (c) project the writer's next turn, providing an object to be introduced in subsequent talk.

6.4 The spatial arrangement of marks on the whiteboard reflects both a conceptual ordering between items and the sequential order of their production

The use of the whiteboard to represent logical relations is a practical, embodied accomplishment. Each next entry onto the board must be organized with reference to the opportunities and limitations provided by previous entries given the physical confines of the available space. At the same time, the necessary juxtaposition of items is a resource for representing meaningful relations among them. The significance of spatial organization among items is to some extent conventionally established (e.g., the list), in other ways dependent on the contingencies of the particular items' production.

6.5 Whiteboards may be delineated into owned territories, or inhabited jointly: Similarly with particular items

Use of the whiteboard varies between more and less exclusive activity by a "scribe" to joint use, and the use of space varies from territoriality (often just on the basis of proximity) to shared access. Territory or items entered by one participant may become joint as others add to or modify them.

6.6 Items entered on the whiteboard may or may not become records of the event

Writing done on the whiteboard may be communicative without being documentary. An extreme case is the "ghost" entry – a gesture at the board that never actually becomes a mark but can nonetheless be referred back to in subsequent talk (Garfinkel and Burns, 1979). Less extreme forms are various cryptic lines, circles and the like that direct attention and accompany talk but are not themselves decipherable. Items can be and often are erased, indicating their status specifically as *not* part of the record, and the status of the talk that produced the item as an aside or digression. Alternatively, an item constructed as illustration may effectively become a document of the talk.

6.7 The whiteboard is a setting for the production and resolution of design dilemmas

Like any practical activity, research and design work encounters both routine and remarkable troubles, the latter becoming objects for reflection and resolution. But in design the dilemmas are not only expected but actively looked for. As a way of proceeding, the designer's task is to make trouble for herself in the form of unsolved problems and unanswered questions. Represented on the board, those problems and questions provide the setting for subsequent actions. Work at the whiteboard thus involves the resolution a series of dilemmas of its own making.

6.8 The whiteboard is embedded in a network of activities

While the whiteboard comprises an unfolding setting for the work at hand, the items on the board also index an horizon of past and future activities. The outcomes of previous actions are reproduced as the basis for what to do now, while what gets done now makes reference to work to be done later. Nonetheless within this network of their own and others' ongoing activities, scientists manage somehow to bound their activities in ways that bring closure each time for this time and place.

7. Conclusion

The situated practice of work at the whiteboard underscores a phenomenon observed elsewhere in social studies of science (Collins, 1985; Knorr-Cetina, 1981, Garfinkel, et al., 1981; Lynch et al., 1983). While scientific reasoning consists in negotiating practical contingencies of shop talk and its technologies, those practices are notably absent from the scientific outcomes and artifacts produced. This absence is not offered by sociologists of science as an irony, but rather as an observation with profound implications for how we understand the status of representations in science and elsewhere: *viz*. we must understand them in relation to, as the product of and resource for, situated practice. Just as instructions presuppose the work of "carrying them out," so representational devices assume the local practice of their production and use. Such situated practice is the taken-for-granted foundation of scientific reasoning.

While the rational artifacts of cognitive scientists' work are programs that run, cognitive scientists' own rationality is an achievement of practices that are only *post hoc* reducible to either general or specific representation. Canonical descriptions do not and cannot capture "the innumerable and singular situations of day to day inquiry" (Lynch et al., 1983:209). The consequence is a disparity between the embodied, contingent rationality of scientists' situated inquiries and the abstract, parameterized constructs of rational behavior represented in computer programs understood to be intelligent. To the extent that cognitive science defines the terms of rational action the disparity is not only theoretically interesting, but has political implications as well. In particular, science studies recommend indifference toward the relation of representation to phenomenon, in favor of a focus on the practices by which representations of phenomena are produced and reproduced. In the case of cognitive science, however, the phenomena are just those things on which our studies take a stand; namely, the organization of practice. In turning to the work of cognitive scientists, therefore, we have a vested interest – not only in the products of cognitive scientists' theorizing but in the adequate rendering of their and others' situated practice.

Notes

1. The study of whiteboard practices is part of a larger project with Randy Trigg to investigate how computer-based technologies might support scientific research practices. "Whiteboards" are just like blackboards but are white, and are written on with colored markers.
2. As Michael Lynch puts it: "Given how easy it is to constitute a docile subject as intentional, it raises the question of how machine intelligence can possibly be extracted from such interactional work" (personal communication).
3. Garfinkel's (1967) original inversion, on which ethnomethodology is founded, has to do with Durkheimian proposals regarding the nature of social facts:

   > Thereby, in contrast to certain versions of Durkheim that teach that the objective reality of social facts is sociology's fundamental principle, the lesson is taken instead, and used as a study policy, that the objective reality of social facts as an ongoing accomplishment of the concerted activities of daily life, with the ordinary, artful ways of that accomplishment being by members known, used and taken for granted, is, for members doing sociology, a fundamental phenomenon. Because, and in the ways it is practical sociology's fundamental phenomenon, it is the prevailing topic for ethnomethodological study. (p. vii)

4. I owe this example to a talk by Terry Winograd. For a wide-ranging critique of the "rationalistic tradition" of cognitive science and alternate proposals for computer design, see Winograd and Flores (1986).
5. I am indebted for this example, and many clarifying discussions of planning, to Randy Trigg.
6. For a recent attempt to develop a computational account of "abstract reasoning as emergent from concrete activity," see Chapman and Agre (1986); and Agre and Chapman (1987b).
7. For an eloquent treatise on the situated structuring of activity, see Lave (in press).
8. For a detailed treatment and the evidence for these conjectures, see Suchman and Trigg (forthcoming).

References

Agre, P., and Chapman, D. (1987a). *What are plans for?* Paper presented for the panel on Representing Plans and Goals, DARPA Planning Workshop, Santa Cruz, CA., MIT Artificial Intelligence Laboratory, Cambridge, MA.

Agre, P., and Chapman, D. (1987b). *Pengi: An implementation of a theory of activity.* Proceedings of the American Association for Artificial Intelligence, Seattle, WA.

Allen, J. (1983). Recognizing intentions from natural language utterances. In M. Brady and R. Berwick (Eds.), *Computational models of discourse*. Cambridge, MA: MIT Press.

Boden, M. (1973). The structure of intentions. *Journal of Theory of Social Behavior* 3:23–46.

Chapman, D., and Agre, P. (1986). Abstract reasoning as emergent from concrete activity. In M. Georgeoff and A. Lansky (Eds.), *Reasoning about actions and plans: Proceedings of the 1986 workshop*. Los Altos, CA: Morgan Kaufmann.

Collins, H. (1985). *Changing order: Replication and induction in scientific practice*. London: Sage.

Collins, H. (1987). Expert systems and the science of knowledge. In W. Bijker, T. Hughes and T. Pinch (Eds.), *The social construction of technological systems*. Cambridge: MA: MIT Press.

Dennett, D. (1978). Brainstorms. Cambridge, MA: MIT Press.

Gardner, H. (1985). *The mind's new science*. New York: Basic Books.

Garfinkel, H. (1967). *Studies in ethnomethodology*. Englewood Cliffs, NJ: Prentice Hall.

Garfinkel, H., and Burns, S. (1979). Lecturing's work of talking introductory sociology, Department of Sociology, UCLA. To appear in *Ethnomethodological studies of work*, Vol. II. London: Routledge and Kegan Paul.

Garfinkel, H., Lynch, M., and Livingston, E. (1981). The work of a discovering science construed with materials from the optically discovered pulsar. *Philosophy of the Social Sciences*, 11(2):131–58.

Grice, H.P. (1975). Logic and conversation. In P. Cole and J. Morgan (Eds.), *Syntax and semantics*, Vol. 3: *Speech Acts*. New York: Academic Press.

Gumperz, J. (1982). The linguistic bases of communicative competence. In D. Tannen (Ed), *Georgetown University roundtable on language and linguistics: Analyzing discourse: text and talk*. Washington, DC: Georgetown University Press.

Jordan, B., and Fuller, N. (1974). On the non-fatal nature of trouble: Sense-making and trouble-managing in *lingua franca* talk. *Semiotica* 13:1–31.

Knorr-Cetina, K. (1981). *The manufacture of knowledge*. Oxford: Pergamon Press.

Latour, B., and Woolgar, S. (1979). *Laboratory life: The social construction of scientific facts*. London and Beverly Hills: Sage.

Lave, J. (1988). *Cognition in practice*. Cambridge, UK: Cambridge University Press.

Livingston, E. (1978). *Mathematicians' work*. Paper presented in the session on Ethnomethodology: Studies of Work, Ninth World Congress of Sociology, Uppsala, Sweden. To appear in Garfinkel, H., *Ethnomethodological studies of work in the discovering sciences*, Vol. II. London: Routledge and Kegan Paul.

Lynch, M. (1981). *Art and artifact in laboratory science*. London: Routledge and Kegan Paul.

Lynch, M., Livingston, E., and Garfinkel, H. (1983). Temporal order in laboratory work. In K. Knorr-Cetina and M. Mulkay (Eds.), *Science observed: Perspectives on the social study of science*. London and Beverly Hills: Sage.

Lynch, M., and Woolgar, S. (1988). Introduction: Sociological Orientations to representational practice in science. *Human Studies* 11(2–3):99–116.

Miller, G., Galanter, E., and Pribram, K. (1960). *Plans and the structure of behavior*. New York: Holt, Rinehard and Winston.

Sacks, H., Schegloff, E., and Jefferson, G. (1974). A simplest systematics for the organization of turn-taking in conversation. *Language* 50(4): 696–735.

Searle, J. (1979). *Speech acts: An essay in the philosophy of language*. Cambridge, UK: Cambridge University Press.

Suchman, L. (1987). *Plans and situated actions: The problem of human-machine communication*. Cambridge, UK: Cambridge University Press.

Suchman, L., and Trigg, R. (in press). Constructing shared conceptual objects: A study of whiteboard practice. In J. Lave and S. Chaiklin (Eds.), *Situation, occasion, and context in activity*. Cambridge, UK: Cambridge University Press.

Tibbetts, P. (1988). Representation and the realist-constructionist controversy. *Human Studies* 11(2–3):117–132.

Turing, A.M. (1950). Computing machinery and intelligence. *Mind* 59 (236):433–61.

Weizenbaum, J. (1983). ELIZA: A computer program for the study of natural language communication between man and machine. *Communications of the ACM, 25th Anniversary Issue* 26(1):23–3 (reprinted from *Communications of the ACM* 29(1):36–45, January 1966).

Winograd, T., and Flores, F. (1986). *Understanding computers and cognition: A new foundation for design*. Norwood, NJ: Ablex.

Woolgar, S. (1983). Irony in the social study of science. In K. Knorr-Cetina and M. Mulkay (Eds.), *Science observed: Perspectives on the social study of science*. London and Beverly Hills: Sage.

Woolgar, S. (1985). Why not a sociology of machines? The case of sociology and artificial intelligence. *Sociology* 19(4):557–572.

Following instructions

RONALD AMERINE
JACK BILMES
Department of Anthropology, University of Hawaii at Manoa, Honolulu, HI 96822, USA

To discover some of the implicit and generally unrecognized cognitive tasks which underlie the achievement of coherent or "accountable" cognitive performances we examined videotapes of a series of science experiments in a third grade classroom. These experiments are part of a commercial "multimedia" science program, "Amazing Adventures."[1] This program is comprised of animated film-strips and illustrated storytexts depicting "Cosmos the Incredible" and his young friends performing extraordinary, seemingly magical feats; these turn out to be based on natural scientific principles which are the subject of student science experiments, conducted in accordance with instructions provided by "Activity Sheets" correlated with the film strips.

Our approach to these data is influenced most directly by the recent work of Harold Garfinkel and his students (Garfinkel, in press; Garfinkel, Lynch and Livingston, 1981; Lynch, Livingston and Garfinkel, 1983). Garfinkel is concerned with the practical contingencies, the "lived work," of accomplishing "naturally accountable" activities, such as forming service queues, following map directions, and making scientific discoveries. In our accounts, both as members and as social scientists, of human activities, we tend to ignore the mundane or seemingly insignificant details of how those activities were actually produced within a specific setting. Garfinkel writes of

1. A copyrighted (1979) product of Nystromg, Division of Carnation Company.

> ... "horizonal" properties of naturally available phenomena [such] as their historicity, their detail, their developing intelligibility, their circumstantiality, their contingent occurrence, and their embedded production. Canonical problems of social order are practical methods for theorizing the contents of everyday activities by furnishing grounds for treating the horizonal properties as irrelevant ...
>
> The expressions, "unremark-able" and "unnotice-able" are hyphenated in referring to practices of such unquestioned efficacy and banality that no motive ordinarily exists, either in commonplace settings or professional inquiries, to make an issue of their methodic character. In the social scientific search for routine, predictable, standardized, and orderly states of affairs in the society, these practices are overlooked, while at the same time their routine, predictable, standardized, and orderly production of worldly matter of fact and conjecture incessantly "works for" the social science inquiry (Garfinkel, in press).

The indexicality, incompleteness, and ambiguity of rules and instructions, and the status of these properties as necessary and essential rather than incidental or remediable, has been a major topic of ethnomethodology from its early development (Garfinkel, 1967:Ch. 1; Wieder, 1970, 1974; Zimmerman, 1970) until the present. The recent work of Friedrich Schrecker (1981) on the progress of a laboratory experiment is of particular relevance in the present context. "... the sheet of lab instructions used by Schrecker in his lab work required of students that they locate the text's instructions and, accordingly, the answers and practical reasoning conveyed by the text's specifications, by turning away from the text and initiating embodied activities on the distinctive surface of the lab bench" (Lynch, Livingston and Garfinkel, 1983). Schrecker, like the children in our study, had to turn a set of instructions into a concrete course of work and face the practical contingencies created thereby. As we shall see, for children, the translation from instructions to performance is particularly hazardous, engendering diverse, unforeseen, and quaint difficulties. The result is not that the children do not learn, but that they learn something rather different from what the "experiment" is designed to teach them.

It is notable that the instructions provided to the children (examples available upon request from author) are not merely instructions – they are also prospective accounts. That is, if the experiment is "successful," if it achieves its projected outcome, the instructions can serve as an account of "what was done," although in any actual performance a great deal more is necessarily done than can be comprised in the instructions. It is only when things go wrong that the details of the course of work require examination in the search for an account of what happened. This brings up another property of instructions: it is possible to imagine a set of instructions with no particular projected outcome. Perhaps one might even want to argue that such things occur in the realm of moral imperatives. In all other cases, though (at least those which we can bring to mind), either the instructions lead to a specified or generally known outcome, or to an outcome known to the writer of the instructions and to be discovered by the person undertaking to follow the instructions. Instructions have a projected outcome, known to the instructor and possibly the instructed as well. This property is not definitive of instructions, but it is crucial to the process of following them and accounting for "what happened," as we shall see in the data that follows.

Garfinkel has demonstrated and investigated the hidden (or, perhaps we should say, all-too-obvious) structure of ordinary activities by introducing anomalies into them. What happens, for instance, when a son behaves like a polite guest in his own home, when a blind man asks for place directions, or when a person wearing inverting lenses tries to sit down in a chair? In the case of Agnes, the transexual, Garfinkel (1967:Ch. 5) found a naturally occurring resource for his investigations of the structure of ordinary activity. Agnes, having been raised as a male, had to teach herself how to be a competent, "naturally accountable" female. Children, all children, are comparable to Agnes, and a comparable resource for social scientists, in that the child is incompetent in the ordinary, taken-for-granted skills of daily life. There could hardly be a more "perspicuous setting" (Garfinkel's phrase) for discovering the unremark-able and unnotice-able practices involved in instruction-following than a setting in which young children are called on to follow a set of instructions.

Even when intended as a guide to a comparatively simple course of action yielding easily describable results, instructions and related explanations presuppose a range of competencies and conventional understandings, without which even the most detailed instructions are meaningless for organizing practical activities. This is particularly evident in those cases where the third-graders we studied lacked some of these skills or understandings, frequently with the effect of transforming the experiments into something quite different from what was envisioned in the instructions.

Courses of action prescribed by instructions vary considerably with respect to degrees of skill and comprehension required to carry them out, just as instructions themselves vary greatly in terms of clarity and completeness. But some of the competencies and understandings to which we refer, and those we are most interested in here, are of such a general nature, that is, they seem so fundamental to successfully following any set of (adequate) instructions, that they may be regarded as constituting the essential competence which enables one to follow instructions *per se*. Put another way, they define what one does in following instructions in general.

Successfully following instructions can be described as constructing a course of action such that, having done this course of action, the instructions will serve as a descriptive account of what has been done, as well as provide a basis for describing the consequences of such action. However, like instructions, this description leaves undefined the practical skills, the embedded activities, and the background knowledge, in other words, the competence by means of which constructing courses of action in accordance with sets of instructions is accomplished. We suggest that, rather than learning "science," the primary cognitive task confronting our subjects in these experiments was that of developing such competence – a competence which, because of its problematic status, becomes explicit by virtue of being a resource for interpreting the children's behavior.

Perhaps the most important of the cognitive skills required for dealing competently with instructions is the ability to grasp at the outset some of the general relationships and possible connections between a projected outcome and a corresponding course of action on the basis of information given in the instructions, and in the case of the experiments discussed here, in the "Reason" or "Explanation." This despite the fact that in didactic experiments the discovering of such relations is envisioned as a *consequence* of following instructions, rather than as a *condition* for doing so. Yet it is only by inferring some sort of pattern that the necessarily incomplete nature of instructions can be developed into a coherent course of practical activity; that unavoidable ambiguities and unforeseen contingencies can be resolved appropriately; that one can distinguish that which is essential from that which is nonessential in the instructions; and that one can decide whether any particular action among the virtual infinitude *not* specified by the instructions might facilitate, interfere with, or prove totally unrelated to the outcome. As we will see, all of these skills are required for competently following instructions, though as a consequence of the reflexivity of a course of action and its outcome they depend largely upon anticipating relationships between these last two factors.

Consider, for example, the instructions for the experiment called "Keeping Dry Under Water." A napkin is to be pushed down into an eight-ounce plastic tumbler and the tumbler then inverted and plunged straight down into a plastic bowl half filled with water. The tumbler is to be held in the water for a second or two and then lifted straight out. The napkin will remain dry. It will be obvious to a competent adult that these instructions include a number of details that are not essential to the experiment. One could achieve the same result by plunging a 10 1/2-ounce soup can with a rag in the bottom into a bathtub three-quarters full of orange juice and keeping it there for an hour of two. Much of the content of these instructions is therefore determined by practical considerations which are irrelevant to the projected outcome. But one cannot presume that a third-grader would know this. And in fact one of the essential instructions, that the tumbler be lifted straight out of the water, was violated several times, resulting in failures to achieve the projected outcome. The apparent reasons for these departures from the instructions further illustrate the implicit competencies which underlie instruction-following. There is nothing in the instruction sheet that tells (or allows one to deduce) what will happen if the tumbler is tipped while under water. Yet it is precisely this knowledge that is required to correctly understand the meaning of the word "straight" in this context. We would not, for instance, say that a ball did not go straight simply because it revolved in flight. Our understanding of the meaning of "straight" in the instructions is informed by our knowledge of what will happen if the cup is tipped. Rather than saying that several children failed to follow the instruction to lift the tumbler straight out of the water, it would be more accurate to say that they failed to follow the instruction as a competent adult would have interpreted that instruction. This appears to reflect also an unforeseen contingency which arose in the course of the experiment: the napkin often fell out of the tumbler, either before placing the latter in the water or upon lifting it out. Thus some of the children who had successfully gotten the cup and napkin into the water subsequently tipped the cup to ensure that the napkin would not fall out when they raised it. Some of the students met this contingency by suggesting that tape be used to hold the napkin in place, a method adopted by several others; but it is interesting to note that many of the children rejected this solution, preferring the challenge of trying to succeed without such assistance. The latter portion of this science lesson therefore evolved into a competitive social activity, students who succeeded without using tape being rewarded with cheers and applause from their classmates.

This denouement is not inconsistent with what we have been saying about instructions. In making a competitive game out of following instructions which, in a very few years, they will find trivial and so easy to carry out as hardly to require conscious thought, these children are demonstrating that the ability to turn instructions into practical activities that achieve predictable outcomes is not yet an implicit, taken-for-granted competence, but a set of skills which they are in the process of developing. So it was not the problem of "air pressure" so much as the problem of constructing a coherent, "successful" course of action out of the experimental instructions with which they became engaged.

Several incidents we observed illustrate the need for recognizing connections between the projected outcome and the ongoing activity in order to avoid more or less random actions which interfere with the experiment. For example, in the case of "Invisible Writing," where students write with a toothpick dipped in salt water and subsequently produce an image by rubbing carbon paper across the residual salt crystals, we observed several children licking the salt off the toothpick before writing with it. Several others, in rubbing their fingers over the paper in order to feel the dried salt crystals, appear to have wiped the salt away. Not surprisingly, this experiment produced few unambiguously successful outcomes. In the experiment entitled "Making Water Wetter," in which dipping a soap-covered finger into the center of a cup of water sprinkled with pepper causes the pepper to move to the edge of the cup, according to the instructions as a consequence of the soap breaking the surface tension of the water, some students produced this effect simply by bouncing their fingers up and down or stirring them around in the water so vigorously as to create waves which pushed the pepper to the outside.

The pattern which inheres in a coherent set of instructions, and which in turn makes such instructions coherent, not only guides actions, but determines perceptions as well, in that it tells one what to look for, what to regard as relevant observations, and what to ignore. Such a channeling of perceptions is necessary not only in order to regulate the practical course of action but to determine if the projected outcome is in fact achieved. Thus competence in dealing with instructions is at the same time a very situated competence in "viewing the world," or "seeing what is there," according to the account of things embodied in the instructions. Because they had not fully developed such a competence, our subjects frequently ascribed significance to observations which a competent adult would regard as irrelevant, "out of frame," or otiose with respect to a coherent "scientific" account of what was being done.

An example of this may be seen in the "Keeping Dry Under Water" experiment. To expedite carrying out this lesson two similar and functionally equivalent pans of water were placed on a table in the center of the room and the students were called on by pairs to try the exercise. Toward the end, when, as related above, this activity had become particularly competitive, one of the children approached a pan but was urged by classmates to use the other one because it was "luckier." We are not sure how this notion came about, although in a pair of trials closely preceding this comment the student using the "unlucky" pan had failed, while the child using the other one had succeeded. At any rate, the student followed this advice and the experiment was successful. Both of the following two children rushed for the "lucky" pan, though the loser settled for the "unlucky" one (and succeeded nevertheless). In the case of the next pair, the second child waited for the first to finish using the "lucky" pan, and then also used it. The "unlucky" pan remained unused thereafter.

In another experiment, the children were instructed to hold a slip of paper just below their mouths and blow across the top of it. The expected result being that the paper would rise due to the reduced pressure of the air moving over it. One of the students was unable at first to produce this effect and a classmate suggested that she was holding the slip of paper with the wrong hand.

In neither case are such observations *by nature* illogical or irrelevant. If a child were having difficulty learning to, say, bat a ball right-handed it would be appropriate to ascertain, perhaps by experimentation, if he were left-handed; and if one were unable to decide which of two brands of automobile to buy, she might reasonably take into account the good or ill fortune of any acquaintance(s) who had recently bought one or the other. But in these science experiments our understanding of the relationship between the practical course of action and its outcome seems to leave no place for "luck" or handedness. Therefore such factors become "noise"; they are outside of the frame of reference defined by the instructions.

This "framing," by which the complexity of the perceivable world is more or less spontaneously organized, is also evident in the decision as to whether an actual outcome sufficiently resembles the projected outcome described in the instructions that the experiment is to be regarded as a "success" or as a "failure." Phenomena often do not lend themselves unambiguously to such discontinuous classifications, but in these instances, it is necessary to order phenomena so as to yield practical classifications in accordance with criteria given in the instructions. Instructions, furthermore, by their very nature lead us to expect that, assuming we have followed them correctly, the projected outcome will occur. Thus our interpretation of outcomes involves expectations not only concerning *what* should occur, but also *that it should* occur. As the following examples illustrate, competence in this regard requires producing conceptual order put out of phenomenal ambiguity without letting prospective accounts of "what is there" preclude alternative, contingent accounts.

In the case of "Invisible Writing" with salt water, for reasons given above many of the children were unable to make anything even approaching legible writing appear, though by vigorously and persistently rubbing the carbon paper over their papers they did produce irregular blotches and streaks. They often tried to persuade themselves and their classmates that these constituted successful outcomes, attempting to show how certain random marks might be interpreted as particular letters. In the case of "Making Water Wetter," when the first student dipped his finger into the water some of the pepper sank while some went to the sides of the cup. One student immediately exclaimed "success!" while another said, "they're going down to the bottom." When the latter statement was amended by the teacher's observation that some (actually, only a few flakes) went to the bottom and some to the sides, consensus was achieved that the projected outcome, *viz.*, "the pepper will move quickly to the outside of the tumbler," had in fact occurred. The students here achieved a competent, "in frame" interpretation of the results, but only after a certain amount of negotiation. It might be argued that they learned something here about the proper seeing of results produced according to instructions.

One of the other students suggested that the experiment would have the same outcome if small pieces of paper were substituted for the pepper, a prediction which most of the children responded to with disbelief, some even with derision. When this was tried, once again some of the pieces sank while others moved to the outside. In this case the overwhelming consensus was that because some of the paper had sunk, the experiment had failed.

In the initial experiment, the authority of the instructions was decisive in classifying the objectively ambiguous results. "What happened," as far as most of the students were concerned, was that the pepper moved to the sides, as predicted; negative instances were (eventually) discounted as irrelevant. But in the improvised experiment (which was in fact the true "experiment"), lacking such authority and at the same time expecting failure, the children conversely refused to see as overriding those instances where the paper moved to the sides and instead classified the outcome in terms of the paper sinking, i.e., as a failure. (The fact that some of the pepper and paper sank might be seen as a powerful demonstration of the principle of surface tension, but it was not envisioned in the instructions. For the students, concerned with "success" and "failure" rather than with the scientific principles that the experiment was ostensibly teaching, the sinking was unexpected and untoward and consequently a sign of failure at the practical activity of instruction-following.)

Idealized notions of science as an abstract, disembodied enterprise are, as we have seen, a poor representation of the actual work of doing science. In addition, science is also conventionally presented as abstracted from the social setting in which it occurs. But, as *The Double Helix* by James Watson vividly documents, science is through and through a social enterprise, penetrated with social considerations, and this is at least equally true for scientific "experiments" done in classroom settings. It is not simply a matter of doing something and seeing the results. The results are classified as "success" and "failure" and thus are laden with social implications. The doing of the experiment and the interpretation of the results come to involve social support, competition, gain and loss of face. The nature of the results is a matter not merely for observation but for negotiation. Although we will not go further into these considerations here, any discussion of the socially defined outcomes of the pepper and paper experiments described above would have to take the social contexts of these experiments into account.

In many instances, as we have seen, there were failures to achieve outcomes predicted in the instructions. In virtually none of these cases was such a failure allowed to pass without at least one of the children offering an explanation. This would seem to reflect a common, if implicit, acceptance of instructions as prospective accounts of how projected outcomes are brought about; the correctness of these accounts remained unquestioned, though their completeness, in the sense of providing all relevant details, was often in doubt. A failure, therefore, might bring into question the completeness but never the correctness of the instructions. The "experiments" did not test the validity of a scientific principle, only the competence of the students at carrying out the instructions. The children were also provided with the occasion to practice a useful social skill – accounting for discrepant outcomes within a framework of unquestioned authority.

In this sense, and unlike in the case of hypothetical experiments, it may be said that rather than learning how to use evidence to reason from controlled conditions, the students were learning the practical skills and imagination involved in rationalizing such evidence, that is, in *ad hoc* speculation concerning violations of or incompleteness in instructions. For example, in an experiment involving the use of liquid dish soap to blow bubbles through plastic straws, a few of the students who were unable to blow bubbles as large as those expected on the basis of the instructions took this as indicating that the brand of soap employed was inferior. The failure to produce a legible message in "Invisible Writing" was said by some to be due to using the wrong kind of paper.

A common feature of the failures to accomplish expected outcomes which we observed was their lack of any real theoretic interest; they were rationalizable in terms of retrospective accounts of practical courses of action, rather than explainable in terms of general principles. The result, as suggested earlier, is not that the children fail to learn, but that they learn something different from what the experiment is intended to teach them. What they learn are, most importantly, the practical and creative skills needed to successfully turn a set of instructions into an accountable course of action, or, if necessary, to account for failure without discrediting the instructions.

We have suggested that dealing competently with the instructions requires not just the apprehension of bare imperatives, but an understanding of general relationships and possible connections between a projected outcome and a corresponding course of action, of which the instructions are indexical. This indicates the reflexivity, or the mutually determinative nature of the course of action and its outcome, in which is grounded the meaningfulness and coherence of a set of instructions. The course of action is determinative of the outcome not only in a physical sense but in that the course of action, as it comes to be formulated in subsequent accounts, makes certain aspects of the outcome noticeable, relevant, and mentionable. The perceived outcome, on the other hand, informs one's perception and account of what the course of action was. The same course of action may be differently described in accordance with what outcome it appears to have produced. This is especially the case when the projected outcome does not materialize and one has to examine one's course of action to see if and how it was consistent with the instructions. In such cases, previously insignificant and irrelevant details may become crucial in an account of the course of action. There is another aspect, though, to this reflexivity: One's sense of the course of action prescribed by the instructions is informed by one's knowledge of the projected outcome, just as one's sense of what will serve to constitute and be essential in such an outcome is informed by the prescribed course of action. It is in this way that the meaningfulness and coherence of instructions is grounded in the perceived relationship between course of action and projected outcome.

As our observations of these third-graders indicate, it is largely by means of achieving competence with respect to the indexical and the reflexive nature of instructions that one becomes able to recognize the essential and unessential features of the accounts embodied in instructions; to fill in the gaps in these accounts, both conceptually and through practical activities; to determine the relevance of particular acts; and to reduce ambiguity by means of practical classifications of phenomena. Although our subjects were in many respects less than competent in these skills, they seemed clearly to possess well-developed senses of "accountability" as an organizing and interpretive principle of practical activities (and their outcomes). By virtue of this sense of accountability as the form according to which meaning is ascribed to actions, and actions are constructed out of meanings, the cognitive skills tapped and developed by elementary science experiments were far less of a "theoretical" (in the usual sense) than of a practical nature.

References

Garfinkel, H. (1967). *Studies in ethnomethodology*. Englewood Cliffs, NJ: Prentice Hall.

Garfinkel, H. (In press). *Forward to ethnomethodological studies* (Vol. 1). London: Routledge and Kegan Paul.

Garfinkel, H., Lynch, M., and Livingston, E. (1981). The work of a discovering science construed with materials from the optically discovered pulsar. *Philosophy of the Social Sciences* 11:131–158.

Lynch, M., Livingston, E., and Garfinkel, H. (1983). Temporal order in laboratory work. In K. Knorr and M. Mulkay (Eds.), *Sciences observed: Perspectives on the social study of science*. London: Sage.

Schrecker, F. (1981). *Doing a chemical experiment: The practices of chemistry students in a student laboratory in quantitative analysis*. Unpublished paper, Department of Sociology, University of California, Los Angeles.

Wieder, D.L. (1970). On meaning by rule. In J.D. Douglas (Ed.), *Understanding everyday life*, 107–135. Chicago: Aldine.

Wieder, D.L. (1974). *Language and social reality*. The Hague: Mouton.

Zimmermann, D.H. (1970). The practicalities of rule use. In J.D. Douglas (Ed.), *Understanding everyday life*, 221–238. Chicago: Aldine.

The dictates of method and policy: Interpretational structures in the representation of scientific work

STEVEN YEARLEY *
Department of Social Studies, Queen's University, Belfast BT7 1NN, Northern Ireland

1. Introduction: Representations of scientific work[1]

Since the middle years of this century many leading scientists in the West have spoken out in support of claims for the autonomy of basic science. Such claims have been bolstered in various ways: by warnings about the political enslavement of science which would follow from restrictions on its autonomy and by essentially functionalist arguments according to which science grows best when freed from external regulation. These defences of the 'Republic of Science' have generally sought to secure for scientists the right to decide for themselves on the distribution within science of forms of state support like research grants (Ronayne, 1984:75–83). Because the demand for autonomy was more or less successfully tied to the supposed requirements of scientific knowledge itself, this strategy for furthering scientists' political interest has been very attractive. It justifies a step which is largely in the profession's self-interest in terms which bear no relation to self-interestedness at all. Whatever scientists as individuals might wish, it could be argued, it is in the nature of science to demand independence from social regulation. It simply needs to

* I should like to express my thanks to participants at the George Sarton Centennial meeting in Gent and at the ESRC-supported workshop on Reflexivity and Discourse Analysis in St. Andrews for their comments on earlier versions of this paper and to this volume's editors for their most helpful recommendations. The research on which this paper is based is supported by an ESRC award (Grant No. A33250031).

be autonomous. An image or representation of scientific work has thus come to be important, as both Gieryn (1983) and Mulkay (1976) have observed, for defences of the scientific 'estate'.

This image of autonomous science has not only had a political career. It has been of considerable scholarly importance too. Many of the general issues about science which have attracted most scholarly and (often) philosophical debate – issues like the rationality of science – have been concerned with the workings of 'pure' scientific knowledge. It is, however, in some ways ironic that such an image of the workings of basic scientific research has come to have considerable political and scholarly significance when work in basic science cannot be considered to be typical or even representative of most scientific activity. It is commonly the case in contemporary science that the direction of research is stipulated by private or governmental bodies in terms of objectives (such as enhanced productivity from a specific process or increased energy efficiency in power generation) which are not consonant with the general image of basic science. Moreover, scientific workers operate within a division of labour, especially concerning how much time they are able to devote to any one project, which is foreign to most university-based basic science. It appears increasingly to be the case that, even in areas which may with some plausibility be regarded as basic science, research is pursued under conditions which do not at all match the idea of a free cognitive 'market' – the ideal form contemplated for autonomous science. For example, it has recently been widely reported that in the UK funding proposals which, a short time ago, would certainly have been accepted because of their high quality are now being refused.[2] At the same time an attempt has been made to promote research in the broad area of information technology through the creation of new university posts and the provision of special funds for research.

Thus, the mechanisms which drive the direction of attention in basic research itself and the processes which lead to the evaluation of projects and funding are openly being affected (and at least in theory 'regulated') by acts of policy. To some extent, therefore, scholarly interest in topics such as the extent to which scientific change is attributable to 'social' or 'rational' elements is simply being overtaken by circumstances. The continuation of scholarly debates over these matters may reflect simple inertia. Meta-scientists may not have come to terms with the transition of science into a commodity in the same way that science planners have (Gibbons and Wittrock, 1985). But an additional factor, to be explored in this paper, is the continuing tendency of scientists themselves to present accounts of scientific work which make slight reference to political actions and economic adjustments. In the use of such a vocabulary, scientists appear to reinforce the image of science as an autonomous enterprise. This presentational style, here termed intellectualist, is used by scientists for various interpretational tasks. Intellectualist accounts are found to persist despite the seemingly far-reaching changes in the policy environment of basic science. This paper will explore the role of this interpretational pattern in scientific culture and suggest ways in which the persistence of such a vocabulary may influence the ability of scientists and science planners to negotiate over science in the policy arena.

2. Telling science the intellectualist way

It is clear that science can be told in an intellectualist manner; accounts which can reasonably be described as intellectualist have been identified on a number of occasions. And these are not simply rationalistic accounts of the growth of science as a whole; they concern the development of individual scientists' careers.[3] Thus, for example, Woolgar (1981) examined the composition of the account provided by the astronomer Hewish in his Nobel lecture; in this lecture Hewish was concerned with outlining his path to the Nobel achievement. In a similar manner, Gilbert (1980) outlined the biographical histories provided by scientists during interviews. Earlier on, Kuhn (1970) had famously described related accounts in textbook discussions of scientific practice.[4]

The excerpt from Hewish's lecture provided by Woolgar supplies a helpful introduction to the features of these accounts. The lecture is composed of assertions about the speaker's scientific development like (1981:249):

> My fascination in using extra-terrestial radio sources for studying the intervening plasma next brought me to the solar corona.

The account is couched in terms of factors immediately associated with the cognitive aspects of science. This is not to say that the account is 'internalistic' in the precise sense that word has acquired in the historiography of science since the speaker invokes not only the acquisition of evidence but also contingent, psychological attributes such as his fascination. But the causal forces at play are presented as cognitive, ideational or intellectual ones. The presentation does imply that science is a self-sufficiently cognitive undertaking. The path of his scientific development is said to be driven by motivations and problem perceptions which are internal to the scientific career.

A related presentational device occurs earlier in the same passage (Woolgar, 1981:249):

> The trail which ultimately led to the first pulsar began in 1948 when I joined Ryle's small research team and became interested in the general problem of the propagation of radiation through irregular transparent media.

In this instance the portrayal of scientific activity is once again internal: it alludes only to opportunities, ambitions and constraints occurring within the speaker's scientific life. Moreover, the account is explicitly organised in a teleological fashion. It presents events in the past as steps or stages in the course of the advance of scientific knowledge. Out of the entire range of potentially significant contextual details concerning the progress of his scientific career, the speaker selects only intellectual stimuli and causes of action. The teleological nature of his account is compounded by one further feature of his talk. He states (1981:249):

> We are all familiar with the twinkling of visible stars and my task was to understand why radio stars also twinkled.

Hewish had a *task*. The presentation of his career can thus be given shape by reference to the carrying out of this task: the solution to the problem of the twinkling radio stars.

Related patterns are recorded by Gilbert (1980). He argues that in interviews scientists tend to treat their interlocutor to a version of science akin to that which they would give to a student. The person's life history, for example, is narrated through a series of academic problems rather than through career opportunities or through external constraints on their discipline. Gilbert (1980:232) records how:

> Once the interview had begun, the scientists tended to respond to my questions ... by reference to the sequence of scientific problems with which they had been concerned. ... Taking their answers to my questions at face value, the impression was given that a scientist's life is defined by and intimately related to a series of discrete 'problems', and an understanding of these is sufficient to understand a research career.

Such a biography typically contains no reference to broadly 'political' matters nor to organisational information which is, as Gilbert notes (1980:232), 'in marked contrast with the response one might expect from, for example, a business executive asked about his career'. Both in regard to the emphasis on the causal impact of cognitive developments and to the exclusion of issues of politics or policy, these narratives resemble the one provided by Hewish. Furthermore, the central role occupied in these biographies by 'problems' is comparable to the position of 'tasks' in Hewish's narrative. The narratives come across as purposive and directional, not messy and circumstantial in the way some histories can be.

Gilbert proposes that sociological interviewers may be treated to accounts of this sort because scientists are inexperienced at talking to social scientists. Lacking a role model for such interactions scientists use the nearest, practised role they know: interaction with a student. It would presumably be a functional requirement of such a role model that unnecessary personal material be excluded and that the scientific facts be allowed to 'speak for themselves'. An alternative interpretation of the nature of these accounts (and one, incidentally, more in line with Gilbert's more recent analyses [Gilbert and Mulkay, 1984]) would be that scientists customarily employ a number of descriptive repertoires which are mobilised to fit different situations. But regardless, for the minute, of which interpretation one places on these accounting practices, the accounts themselves may be generally characterised as intellectualist.

Lastly, Kuhn (1970:138) seems to be encountering a related phenomenon when he writes that, in textbooks:

> Parly by selection and partly by distortion the scientists of earlier ages are implicitly represented as having worked upon the same set of fixed problems and in accordance with the same set of fixed canons that the most recent revolution in scientific theory and method has made seem scientific.

At one level Kuhn is here drawing attention to the ubiquitous tendency for scientists to interpret the past in terms of the current conception of natural reality (see Latour, 1980). But in his allusion to scientists being presented as 'having worked upon the same set of fixed problems' Kuhn also identifies a particular interpretative pattern in which scientific developments are related to specific *problems* or *questions*. Such a pattern, which is well brought out in Kuhn's discussion of Dalton, mirrors the kind of account that has been detected from the work of Woolgar and Gilbert. In the case of Dalton, Kuhn states (1970:139):

> All three of Dalton's incompatible accounts of the development of his chemical atomism make it appear that he was interested from an early date in *just those chemical problems* of combining proportions that he was later famous for having solved. Actually those problems seem only to have occurred to him with their solutions, and then not until his own creative work was very nearly complete. (Emphasis added.)

It appears that Dalton related the history of his theoretical position as the figuring out of a solution to a pre-existing problem. The story is told not simply with the general benefit of hindsight but with the specific benefit of being able to identify objectives and problems which are said to have been pursued and solved. The account is erected upon the proposition that these problems were there all along; history is then the unfolding of the solution.

From Kuhn's position, however, 'those problems seem only to have occurred' to Dalton at the same time as their potential solution was available.

In one sense the existence and usage of these accounts appear easily understood: such accounts offer versions of scientific experience which are adapted to the context in which they occur (although this is not to imply that there may not be significant interest in the particular construction and operation of these versions). Hewish's tale, for example, is arranged for the humble celebration of the award of great scientific honour (see Mulkay, 1984). But the very existence of such 'histories' and biographies and their prevalence in certain contexts indicate that it is somehow possible, indeed viable, to offer intellectualist interpretations of areas of scientific activity which must have been greatly affected by policy-type influences.[5] In fact, the possibility of intellectualist accounts is also reflected (as Kuhn, too, 1970:138, notes) in the professional practice of internalist historians of science. How then are these partial narratives generated and what part do they play in scientists' interpretative procedures?

3. Scientists' accounts of the development of research

The accounts considered so far have been offered in circumstances where scientists are telling science to outsiders or specialised audiences. However, studies of routine scientific discourse (arguments presented in formal papers and reports), dating at least from Medawar's work (1963), indicate that in these conditions too partial narratives are generated. Accounts are composed through the selective, but generally consistent, presentation of scientific activity. Medawar (1963:377) makes the point that scientific papers commonly begin with a fictionalised history of the investigation. The paper is presented in such a way as to pose a problem which the author is able to resolve, even though the problem could not have been formulated in such a manner at the start of the research which the paper is reporting.

Such a presentational strategy occurs not only in the written, formal context but also in the presentation of addresses at scientific conferences – whether relatively formal in style (as in the first example) or not (as in the second):[6]

I And the question that had been previously posed of why did it take life so long to arise on earth suddenly reverses itself and you can ask a question such as 'how could life have arisen so early? '

II ((to a slide showing the plants in a field where sheep are grazing)) ...
[we focus on] a field that is of one hectare in which a number of research students have spent their three-year periods lying on their stomachs counting individual plants, individual leaves – focusing their attention in what I could call a Darwinian fashion on the behaviour of individual plants, individual parts of plants and looking at the hazards that are involved in their lives, trying in a way to decide which are the hazards which enter into the life of an individual. ((new slide)) This, in fact, is a somewhat surprising hazard, perhaps to many of you; this is the hazard of being soused in urine from one of the Cambridge sheep.

In addition to displaying the tendency for presentations of scientific work to be organised as though that work had been undertaken in order to answer a pre-given question, these cases both indicate the partiality of such accounts. In the first case, work on palaeontology is presented as a response to a question even though it is then acknowledged that the results of the work are perceived as having *changed the question itself.* In the second instance, the speaker is humorously trying to make the researchers' activities appear bizarre and open to many (possibly risible) interpretations; yet their three-year undertaking and the numerous interests that the ecologists might have had in studying the grazing land are resolved onto the issue of deciding 'which are the hazards which enter into the life of an individual'. For these speakers' purposes – which presumably differ from Gilbert's interviewees' requirements and from the needs of scientific biographies – scientific work is once again set out in relation to putative objectives and pre-existing problems even though the stories could evidently have been related in different terms.

No doubt, setting the presentation out in this way helps to make the audience or readership aware of the kinds of claim and the type of argument which are going to be put forward.

Such a presentational device in scientific addresses may even have further consequences such as the direction of attention away from competing or conflicting claims. But it must be recognised that such a portrayal does require a distinctive and partial presentation of the past. Medawar would go so far as to term this a falsification of the past. But it is not just a simple misrepresentation. This portrayal demands the ascription of a purpose or objective onto the programme, an ascription which can only be achieved with hindsight. A 'falsification' of this sort can only be carried out after the event. After the event such a reconstruction may indeed appear convincing and valid. Thus an intellectualist account of one's research appears to be an *ex post facto* construction: a construction which can serve to displace or exclude other factors or circumstances which might have been invoked as the reason or cause of having undertaken one's research.

4. Informal descriptions of research programmes

A closer understanding of the presentational devices employed in intellectualist accounts can be gained from more intensive study of their occurrence. In this section material derived from a series of interviews with scientists working on Precambrian palaeobiology (the study of earth's earliest life forms and their evolution) will be used. In dealing with this interview material a methodology helpfully outlined recently by Potter and Mulkay (1985) has been used. These authors propose that interviews should be valued at least as much for the displays of reasoning which they can evince from interviewees as for the information the subjects offer. As these authors suggest (1985:268), interviews may be adopted:

> as a technique for generating interpretative work on the part of participants, with the aim of identifying the kinds of interpretative repertoires and interpretative methods used by participants and with the goal of understanding how their interpretative practices vary in accordance with changes in interactional context.

In the course of interview talk respondents generally display the range of techniques they use in accounting for their social experiences. Interviewed scientists may, for example, indicate how they make sense of conflicting beliefs in science through an asymmetrical interpretation of the nature of false and correct beliefs (Gilbert and Mulkay, 1984:67–70). Sometimes in interview conversations scientists may be faced with interpretative tasks which they do not ordinarily encounter. On such occasions the structures of their routine interpretational procedures may be made particularly clear: competing interpretational devices may appear through ambiguities, hesitations and self-corrections. Accordingly, interview material can be expected to shed light on the operation of the interpretative device of interest here, the *ex post facto* construction of research paths.

When interviewed scientists describe their career development, as Gilbert noted, they often put forward intellectualist histories. This tendency is apparent in the following statement concerning the development of the respondent's interest in fossil taxonomies (transcript notation is used to indicate cut-off words, and non-lexical tokens):

> III S: em, after dis-after discovering that that completely turned a lot of the accepted practice of classification and nomenclature of the fossils upside down and so I've had to review all the classification and the nomenclature ...

The impulsion to further scientific work is here presented as deriving entirely from the problems and puzzles set by his fossil finds. The early discovery meant that he then 'had to' devote his research to the revision of the scheme for classifying fossils.

However, when questions are pursued in the course of the interview the respondent accepts that there are gross, external influences on the conduct and shaping of his research:

> IV ((concerning a computerised classification system))
> S: Well I've given it up, I just couldn't keep up with the literature because I had to actually punch all this data onto cards.

I: yes

S: and I just haven't got the time at the moment and there's no money for computer resources either and I haven't the time to go chasing you know elusive grants now.

V ((concerning his post))

I: I would like I mean particularly wanted to be based in in the Botany department

S: yes

I: so that I could make contact with botanists but the powers that be put me in Geology

S: ha

I: although I actually work in Botany most of the time now and well there's a shortage of funds as well em. I asked for twelve hundred a year for instrument time and got four hundred so I can't use the microscope as much as I w' would have liked and that's another, another factor.

Even though such remarks are relatively common and despite the apparently great significance that, for example, a shortage of instrument funding would hold for this scientist, mention of these events does not seem to inhibit his ability to organise an account of his research activities in intellectualist terms.

Straight after his discussion of the computer classification system he comments:

VI S: [in earlier work] I developed, what I wanted to do ah I found that all these spherical microfossils'd very distinctive surface textures and

I: yeah

S: what I did was to look at them in the SEM and TEM the er light microscope and compare the two and that is in fact the only efficient way of using the SEM

((continues with technical details)

it's it's too difficult to interpret what you see in the SEM alone. You need to use it in conjunction and *I*

> *wanted to take this one step further* because I thought that the different fold patterns which'd which had been used by er well th'
> ((criticises the work and classificatory system of foreign scientists))
> I thought that the fold pattern of a sphere might be controlled by the structure of the wall ...
> ((emphasis added))

Thus, despite the admitted disruption and the time expended on the classification of fossils (passage IV), this scientist is able to present the task he is now about to turn to as taking earlier work 'one step further'. The senseful connection between this new work and the old is depicted as virtually a logical continuity. Yet the continuity can be made out in different ways: the 'distinctive surface textures' discussed at the beginning of this passage can be described as necessitating a revision of the fossil nomenclature (as in passage III) or as leading to new work on the nature of the microfossils' cell walls (as in the last line of VI).

In a recent article Ziman (1985) has stressed that scientific research is an intentional activity. By saying this, Ziman seeks to emphasise that research trajectories often comprise long-term, planned activities. Yet the indication from this passage (VI) is that the long-term connections between aspects of research activity can be made in a variety of ways. From any history of research a number of coherent, internal research plans can apparently be made out so that Ziman's observation may best be understood as referring to scientists' practices for representing the temporal coherence of their research and not to any once-and-for-all facet of scientific life. Indeed, this tendency to stress intellectual connections is indicated by the fact that when matters of resources and external influences do appear in a positive light they are commonly associated with comments such as 'It just so happened that' scientist Y had some money from the Natural Environment Research Council so that some course of research, conceived all along, could be pursued.

The question of how these various interpretative procedures may relate to each other can be examined with respect to two further rather extensive quotations. In these excerpts the respondent puts forward more complex versions of his career and the interpretative procedures can be tracked in greater detail. The first excerpt relates to a discussion of the scientist's interest in Precambrian microfossils (line numbers are given for later reference):

VII S: and th- when I applied for the ((source of funding)) the the project that I put forward was partly to well the principal object of it was to use this technique to study ulstrastructure of
5 Precambrian fossils to try to find out what they were
I: yes
S: I had to angle it that way because it was the ((name of job)) in biology
10 I: er
S: so I had to make it biological
I: yes
S: and so I I emphasised that it was to study
[what they what they may be
15 I: [bio yeah
S: in comparison to the ultrastructure of living um things and to um review the classification of Precambrian microfossils so that we could arrive at some consensus on
20 Precambrian evolution
I: yes
S: so that erm the importance of this to biology and all that
I: erm
25 S: so what I was taken on to do was the literature work which I've been carrying on with
I: yes
S: and applying these techniques ...

The first and most straightforward point which emerges from this excerpt is that the scientist could have described his research project as contributing to different areas of science. He claims to have recognised this flexibility and to have chosen the depiction which maximised his eligibility for the post. His comments thus provide some elementary support for the idea that scientists are able to 'dress up' their work in a number of ways to appeal to various funding agencies. However, as was noted above, interviews may be more important for what they reveal about people's interpretative skills than for disclosures about their actions. Information of this sort comes from the second half of the excerpt where, after the comment 'so I had to make it biological', the interviewer voices agreement six times whilst the respondent continues to speak. Whether the interviewer's remarks constitute acknowledgements of understanding or whether they function as 'continuers' – eliciting further elaboration – cannot be definitively stated. In either case, the initial point about the redesignation of the research as biological is elaborated in successive utterances. The respondent formulates in various ways the purpose of his research from a biological point of view and these formulations vary from the rather precise (pursuing the literature search) to the extremely general (establishing the importance of Precambrian life for the discipline of biology).

This scientist is clearly able to put forward in rapid succession a number of versions of the purpose or objective of his research. They might be approximately characterised as:

1. Lines 11 & 13: make his research more biological or bring out its biological implications (with an eye to employment) by
2. Lines 14, 16 & 17: comparing fossil and present-day micro-organisms' ultrastructure.
3. Lines 17 & 18: review the classification of certain Precambrian fossils.
4. Lines 19 & 20: by establishing what sorts of thing were alive in the Precambrian, come to consensus on Precambrian evolution.
5. Line 22: thereby advance the state of biology as a whole.
6. Lines 25 & 26: carry on with the literature review to see what Precambrian fossils had been described and what they had been called.
7. Line 28: pursue the microscopy techniques described in passage VI above.

None of these need be regarded as *the* reason or purpose behind his work. They are nonetheless available to be invoked under appropriate circumstances. The local structure of the interview has here encouraged the respondent to produce a series of them. It is however easy to imagine situations (the opening of conference presentations for example) where one or other of these would be deemed the most appropriate.

This view of the respondent's interpretative pattern is supported in a second excerpt where he is discussing the appeal that his work holds for him:

VIII S: well my personal interest is in the, the thing that really intrigues me is the technical side of actually extracting these things in good condition and the techniques you use to study them I mean you know that's the part of it that I enjoy

I: hmm

S: the laboratory work 'n working out what the structure of these things is um the angle that excites me is that these are micro-organisms which have been preserved for a thousand million years. That's the aspect of it that I find exciting. But in er its implication I- I think the biologists neglected palaeontological evidence which is actually hard evidence.

I: uhhuh

S: and eventually I would like to get back into what you might call microbiological publishing and say: look we've got data here that provide constraints for the biochemical evolutionary pathways that the biochemists have dreamed up.

((Goes on to list some follies of biochemists)) ...

In this second excerpt the scientist moves rapidly from a personalised account of his research and adopts a version invoking disciplinary goals and objectives. Two personal narratives, one about the excitement of the technical challenge and one concerning the sheer fascination of immensely antique life forms, are supplanted by a presentation in terms of the revalidation of evidence overlooked by biochemists. Again, there is available a range of potential accounts of why the research gets done and why research takes one direction or another.

This passage provides evidence about one further aspect of the teleological, intellectualist interpretative pattern. This scientist's comments about what he would like to do 'eventually' indicate that research activity is presented as a rolling, purposeful undertaking. Although the past is shown as issuing in the present, the present is also formulated in anticipation of future goals. Thus one's work can be regarded as both the fulfilment of past aims and as one step in the attainment of future objectives. The intentionality of the activity is stressed but the actual specification of intent is changing and multivalued.

Thus in addition to what the respondent himself says about the flexibility of descriptions of research, his interpretative work in the course of the interview displays that flexibility. And this display allows a more precise characterisation of the nature of that flexibility than his testimony alone. This flexibility is not just a matter of the ability to redescribe one's work as zoology or geology or some other neighbouring discipline. Instead it consists in the ability to formulate a great variety of objectives and problem solutions to which one's work can be said to be directed. These objectives may vary from the personal to the disciplinary; they may be small-scale or projected over many years. They all, however, depend on making 'intellectualist' links between the description of one's work and one's claimed objectives.

5. Conclusion

A number of points arising from the preceding analysis can now be drawn together. As Gilbert indicated, scientists can frequently offer accounts of their careers which present careers as successions of problems and their solutions. In this paper we have seen that it is a special property of such accounts not only that they are partial (for all stories are partial in some way) but that they are characterised by reference to objectives and purposes which can only have been constructed in retrospect. Allusions to external conditions such as the availability of funding, the organisation of research, and personal problems can occur in such accounts but these are presented as circumstantial. Such matters are treated as obstacles to, or fortuitous promoters of, the research path, but the overall point of the research undertaking is characterised in intellectualist terms. However, scientists seem able to offer a number of different versions of just what that intellectual aim or objective is. On some occasions, for example, the aim is said to be narrowly disciplinary; sometimes it is described as methodological. But it may also be connected to the general advance of a whole range of disciplines (see Gilbert, 1976). The fact that in interviews scientists are able to offer a series of possible goals in rapid succession suggests that no one of these accounts is uniquely the reason for the research. They are all plausible reconstructions of what the purpose may have been.

The significance of these results is twofold. First, the variability in the formulation of the objectives of research mirrors the diversity of accounting procedures detected in other studies of scientific discourse. It complements these studies by examining scientists' use of their own research histories and provides indirect support for the recommendation that interviews should be valued as occasions for disclosing respondents' interpretative procedures as well as for eliciting informative reports.

Secondly, this analysis of the ability of scientists to offer a variety of depictions of the purpose of their work may have implications for questions of science policy. At one level, the way in which a great number of circumstantial events can be excluded from the histories scientists offer means that, in retrospect, policy influences can often be omitted from most accounts of scientific development. Since many possible intellectualist histories can be formulated, one will search in vain in many scientific accounts for 'evidence' that science did not proceed along an internally-stipulated path. Accordingly, attempts to estimate the extent to which policy decisions are realised in scientific work may be confounded by the retrospective interpretations which scientists routinely generate. Similarly, when making funding applications scientists are generally required to state the objectives of their research. Given that the range of potentially mentionable objectives can be so various, such statements are bound to be flexible and open to revision (Myers, 1985; Yearley, 1984). Yet policy initiatives are customarily organised around some favoured objective or research goal. In the past policy analysts have remarked on how difficult it may be to translate policy objectives into scientific practice. This has often been said to arise because the connection of policy to research is inherently unpredictable and unsystematic, or because research is creative and open-ended. As Salomon (1973:100) expresses it:

> In the case of science, however, the planning effort is subject to this special constraint: that it must set itself the unforeseeable as an object.

Instead of arising principally from the unpredictable nature of science this 'planning constraint' may be attributable to the great interpretative flexibility with which scientists are able to link their work to the general objectives which are to be found in policy guidelines. It would appear therefore that a greater understanding of these interpretative procedures would throw new light on some long-standing problems in science policy implementation.

Notes

1. See, for example, front-page reports in the *Times Higher Education Supplement* on 25 May 1984; 10 August 1984; and 30 August 1985.
2. Thus 'intellectualist' accounts are not identical with the well known 'empiricist repertoire' described by Gilbert and Mulkay (1984:55–58).
3. As the allusion to the work of Kuhn also implies, such accounts occur in historical contexts too; see Oldroyd (1980) and Yearley (1985).
4. I say policy-type since such considerations applied, I take it, to science under patronage as well as under states' policies.
5. These presentations were recorded at the Darwin Centenary Conference, Darwin College, Cambridge 1982. These examples were chosen to represent different degrees of formality.

References

Gibbons, M., and Wittrock, D., Eds. (1985). *Science as a commodity*. Harlow: Longman.

Gieryn, T. (1983). Boundary work and the demarcation of science from non-science. *American Sociological Review* 48:781–795.

Gilbert, G.N. (1976). The transformation of research findings into scientific knowledge. *Social Studies of Science* 6:281–306.

Gilbert, G.N. (1980). Being interviewed: A role analysis. *Social Science Information* 19:227–236.

Gilbert, G.N., and Mulkay, M. (1984). *Opening Pandora's box*. Cambridge: Cambridge University Press.

Kuhn, T.S. (1970). *The structure of scientific revolutions*. Chicago: University of Chicago Press.

Latour, B. (1980). The three little dinosaurs or a sociologist's nightmare. *Fundamenta Scientiae* 1:79–85.

Medawar, P. (1963). Is the scientific paper a fraud? *The Listener* (12 September):377–378.

Mulkay, M.J. (1976). Norms and ideology. *Social Science Information* 15: 637–656.

Mulkay, M.J. (1984). The ultimate compliment. *Sociology* 18:531–549.

Myers, G. (1985). The social construction of two biologists' proposals. *Written Communication* 2:219–245.

Oldroyd, D.R. (1980). Sir Archibald Geikie (1835–1924), geologist, romantic, aesthete, and historian of geology. *Annals of Science* 37:441–462.

Potter, J., and Mulkay, M.J. (1985). Scientists' interview talk. In M. Brenner, J. Brown and D. Canter (Eds.), *The research interview*. New York: Academic Press.

Ronayne, J. (1984). *Science in government*. London: Edward Arnold.

Salomon, J.J. (1973). *Science and politics*. London: Macmillan.

Woolgar, S. (1981). Discovery: Logic and sequence in a scientific text. In K.D. Knorr, R. Krohn and R. Whitley (Eds.), *The social process of scientific investigation*. Dordrecht: Reidel.

Yearley, S. (1984). Analysing science and analysing scientific discourse: On the argumentative strategy of scientists in the public realm. *Zeitschrift für Wissenschaftsforschung* 3:29–37.

Yearley, S. (1985). Vocabularies of freedom and resentment. *Social Studies of Science* 15:99–126.

Ziman, J. (1985). Highlights from the report on 'Science studies and science policy'. *EASST Newsletter* 4:15–19.

Index